CONTENTS

		PAGE
Foreword		vii
Preface		xv
By Way of Explanation		xxxiii

PART ONE
JAMES WHITE'S RAFT JOURNEY OF 1867

I	The Basis of a Legend	3
II	Dr. Parry, Legend Maker	16
III	I Interview the Hero of the Legend	36
IV	Raft Journeys — Imaginary and Real	70

PART TWO
THE AFFAIR AT SEPARATION RAPIDS

I	Major Powell as Historian	97
II	William Hawkins' Story	138
III	Jack Sumner's Account	164
IV	Major Powell's Version Refuted	214

APPENDIX

Commentary on PART ONE, Otis R. Marston	233
Commentary on PART TWO, Martin J. Anderson	253

Reproduction of the Original

COLORADO RIVER CONTROVERSIES

By
Robert Brewster Stanton

With Commentaries
Written For This Edition
By
Otis R. Marston
and
Martin J. Anderson

WESTWATER BOOKS

1982

This edition contains an unabridged and unaltered reproduction of the original work published by Dodd, Mead & Company in 1932.

International Standard Book Number: 0-916370-09-7
Library of Congress Catalog Card Number: 82-60295

Published 1982 by WESTWATER BOOKS
A division of Belknap Photographic Services, Inc.
Box 365, Boulder City, Nevada 89005

©1982 by Westwater Books. Copyright is claimed only in the appendix of the present work. All rights reserved.
Second Printing 1987.

Manufactured in the United States of America by A to Z Printing Company, Riverside, California

Front cover: James White's Raft, *sketch by Paul Nyeland, courtesy Harolds Club, Reno, Nevada*

COLORADO RIVER CONTROVERSIES

(Photograph taken in 1915)

ROBERT BREWSTER STANTON
(1846–1922)

COLORADO RIVER CONTROVERSIES

BY
ROBERT BREWSTER STANTON
(1846—1922)

EDITED BY
JAMES M. CHALFANT

FOREWORD BY
JULIUS F. STONE
AUTHOR OF "CANYON COUNTRY"

Illustrated

DODD, MEAD & COMPANY
NEW YORK   1932

Copyright, 1932
By JAMES M. CHALFANT

ALL RIGHTS RESERVED
NO PART OF THIS BOOK MAY BE REPRODUCED IN ANY FORM
WITHOUT PERMISSION IN WRITING FROM THE AUTHOR

PRINTED IN THE UNITED STATES OF AMERICA
BY THE VAIL-BALLOU PRESS, INC., BINGHAMTON, N. Y.

*To All Truthful
Colorado River Voyagers*

FOREWORD

SINCE he was thorough-going by nature as well as by education, it is hardly to be wondered at that Robert Brewster Stanton, once his attention was centered upon any subject, should have followed it to a conclusion, sparing neither time nor possible expense. Nor is it at all remarkable that after his personal experiences in the canyons of the Colorado he should have taken the keenest interest in any Colorado River problem or controversy.

Unavoidably he learned of the James White story and because of his first-hand knowledge, the utter impossibility of such a journey as White claimed he had made at once convinced him that White either was another Munchausen or that there was some other and more rational explanation. He therefore set himself to the task of finding out the truth. That he succeeded in his quest no reasonable person, after reading the evidence gathered by Stanton and here presented, can longer doubt.

A personal statement from me may seem gratuitous, but I venture to affirm without reservation that no one who has traversed the turbulent waters of Cataract, Marble, and Grand Canyons will admit there is even the remotest possibility that anyone could pass safely through those formidable gorges *on a raft*. Despite this, White's story still finds champions, but only, however, among those who, having no definite knowledge of their own, seem to find a dubious satisfaction in excluding reason from their mental processes.

Of more vital importance than the White myth is the question as to what actually occurred among the members of the first Powell expedition which left Green River, Wyoming, May 24, 1869, and reached Separation Rapid on September 28 following.

Major Powell places upon the three men, O. G. Howland, Seneca Howland, and William Dunn, the responsibility for their leaving the party, but fails to mention any causes for possible dissatisfaction they may have had which finally culminated in their action at that point. By implication at least he charges them with cowardice. Again Stanton has gathered all factual evidence available and presented it impartially, as a historian in search of truth. This he has done so thoroughly that any word or comment from me would be superfluous were it not for the fact that I also knew Jack Sumner very well and that he had unhesitatingly verified to me the incidents set out in his written statement to Stanton. This, in the interest of fairness, I cannot withhold.

Had there been no disagreements among members of the first expedition it is strange indeed that, facing the known difficulties to be encountered, Powell should have chosen an entirely new party for his second undertaking. I asked Sumner about this and he said *he*

FOREWORD ix.

certainly would not have gone with the Major again under any circumstances, and furthermore that, so far as he knew, every other member of the first party felt the same. Doubtless Powell was also of the same mind. Possibly the obscuring veil of facts he did not wish to publish prevented him from seeing the inconsistency of mentioning only his first expedition, yet using without hesitation incidents that unquestionably occurred on the second expedition in the years 1871-72. This he could not have done inadvertently.

That the authorities having in charge the erection of the monument to Powell and his companions on their memorable journey deliberately chose to falsify history by omitting the names of O. G. Howland, Seneca Howland, and William Dunn from the bronze tablet merely proves their utter unfitness for the job. For that, more than anything else, evidences Powell's real attitude toward three men who had accompanied him throughout nine hundred and sixty miles of the total distance of one thousand and forty-three miles that he himself covered.

In the introduction to the 1926 edition of *A Canyon Voyage,* F. S. Dellenbaugh refers to the bronze tablet on the Powell monument in the following words:

"There has been some adverse criticism of the omission but it seems very clear that three men who refused to finish the journey and hampered the progress and success of the undertaking by backing out at a critical moment deserve no honorable mention. They are fully and justly recorded in the Report and in other places."

I hardly agree with Dellenbaugh that Dunn and the Howlands are recorded "fully and justly" in the Report. But even if they were, the fact remains that for every one who reads the Major's book, never a book of general popular distribution in the first place, published in 1875 and naturally long out of print by this time, thousands of travelers and sightseers read the misleading inscription on the monument.

Still, the attitude of the government as shown in the matter of the Powell monument and in Congress' implied sanction of the nonsensical claims for James White is by no means unprecedented. There is at least one very glaring example of deliberate connivance on the part of Washington in the distorting of actual history to suit personal interests and desires. I refer to the effort to take away from the Wright brothers the credit for the invention of a really practicable airplane.

In 1914 officials of the Smithsonian Institution (created by act of Congress in 1846, and governed by a Board of Regents consisting of the Vice President of the United States, the Chief Justice, three members of the United States Senate, three members of the House of Representatives, and six citizens of the United States appointed by joint resolution of Congress) had the Langley plane of 1903 tested by Glen

FOREWORD xi

Curtiss, at the time a defendant in a suit brought against him by the Wright brothers for infringement on their patents. A test was being made at Hammondsport, N. Y., ostensibly to determine whether the original Langley plane was capable of flight. Actually, the test was not made with the machine as designed and built by Langley, nor with an exact copy of it.

Glen Curtiss made a great number of changes in the machinery, some of them of fundamental importance, before it was flown. Many of these alterations were carefully noted and recorded at the time by representatives of the Wright brothers. And yet no candid and complete statement of these changes has ever been made to the public by the officials of the Smithsonian.

On the contrary, every effort was made to preserve the erroneous impression already widespread that the Langley machine was capable of flying. On page 47 of the Annual Report of the U. S. National Museum for 1914 one finds this unblushing assertion: "Owing to a defect in the launching apparatus, the two attempts to fly the large machine during Dr. Langley's life proved futile, but in June last, without modification, successful flights were made at Hammondsport, N. Y."

For years the Smithsonian authorities exhibited the Langley machine as the first airplane capable of carrying a man in sustained flight. Because of pressure brought to bear upon them they have since altered the label on two different occasions, so that it now reads:

FOREWORD

LANGLEY AERODROME
THE ORIGINAL SAMUEL PIERPONT LANGLEY
FLYING MACHINE OF 1903 RESTORED
Deposited by
The Smithsonian Institution

However, the Smithsonian Institution has sought throughout the whole affair to make only a proud apology to the Wright brothers—amounting to a defense of the conduct of certain of its officials in the past rather than a frank acknowledgement of their attempt to pervert history. The failure of the Smithsonian to correct some of these misrepresentations explains the fact that the first really successful heavier-than-air craft, the plane in which Orville Wright made history at Kitty Hawk, North Carolina, on December 17, 1903, is preserved in Kensington Museum in England, and not in the Smithsonian Institution.

Time does not necessarily bring progress in everything. Consider the Aztecs of old Mexico, to whom a truthful record meant much. In his recent book, "Mexico," Stuart Chase climaxes his summing up of the high points of Aztec civilization in these words: ". . . and the writers of untrue books of history it put to death." We can hardly go as far as the Aztecs in this matter. But certainly books like the present volume, openly and frankly militating against the falsifying of history, are heartily to be commended.

Colorado River Controversies fills a place long vacant in the particular phase of our own history with

FOREWORD   xiii

which it deals. To Robert Brewster Stanton particularly, for his tireless efforts in getting at the facts of these vexed questions, and incidentally to James M. Chalfant, for the able editing of his materials for publication, Colorado River history owes a large debt.

JULIUS F. STONE.

Columbus, Ohio,
May, 1932.

PREFACE

Historians ought to be precise, faithful, and unprejudiced; and neither interest nor fear, hatred nor affection should make them swerve from the way of truth.

—Cervantes

THERE have always been two bitterly contested points in the history of the exploration and navigation of the Colorado River. The first relates to the alleged journey of one James White, trapper and prospector, through the canyons on a raft in 1867, two years prior to the first expedition of Major J. W. Powell. The second concerns the circumstances under which O. G. Howland, Seneca Howland, and William Dunn, members of Powell's first expedition of 1869, left the party at the point later named Separation Rapid and climbed out over the canyon rim in an attempt to reach a settlement, only to be slain shortly thereafter by some Shewit Ute Indians.

Though it is in fact an effort to preserve in print the results of a great deal of effort expended on a particular phase of the history of the West, I suppose it is almost a foregone conclusion that the present volume will be classed as another "debunking" book. True, the central figures, James White and Major John Wesley Powell, lose in it a deal of undeserved lustre.

If James White as a man comes off somewhat better than Major Powell, it will be because it was others who enlarged amazingly upon his story, originally told by him quite simply and honestly, whereas the Major was his own historian, but unfortunately an historian who chose to dress up the facts a bit in his own favor, even at the cost of placing the stigma of cowardice and desertion upon certain of his companions in adventure.

When historians fail "to be precise, faithful, and unprejudiced" and when their accounts are generally accepted as true, with the result that, on the one hand, full and proper credit is not given to those who really merit it, or, on the other hand, men's names are unjustly dishonored, there is always the possibility that some future historian will refute their most plausible assertions. It would seem, in fact, the duty of him who is in possession of the facts to let the truth be known, whatever traditions may be shattered in the process.

. Of all those who through the years have interested themselves in attempting to unravel the twisted threads of Colorado River history, surely none has been more thorough and indefatigable than Robert Brewster Stanton, explorer and navigator of the great river in his own right and for many years an assiduous student of everything pertaining to the history of its discovery and navigation.

Robert Brewster Stanton was born in Woodville,

Mississippi, in 1846. His father, Robert Livingston Stanton, a Presbyterian minister and a former President of Miami University at Oxford, Ohio, was an intimate friend and advisor of Abraham Lincoln. He was graduated from Miami with the B.A. degree in 1871. Seven years later Miami conferred upon him an M.A., and somewhat later the degree of C.E. Soon after leaving college Stanton became engaged in railroad survey work.

In 1889 he was employed by the newly-organized Denver, Colorado Canyon, and Pacific Railway Company as chief engineer of an expedition being sent out to make a preliminary survey of the canyons of the Colorado from Green River, Utah, to the Gulf of California, for the purpose of determining the feasibility of building a railroad on a low uniform grade through the canyons. At the time fuel was relatively scarce and expensive in the Southwest, a circumstance which suggested to one Frank M. Brown that a railroad through the canyons to compete with the transcontinental railroads already operating would pay dividends on the business of carrying Colorado coal alone, to say nothing of other potential passenger traffic each way.

An active, effective promoter, Brown succeeded in convincing a group of business men of the value of his idea, with the result that the Denver, Colorado Canyon, and Pacific Railroad Company was formed and Brown named its first president. The Brown project, however, seemed ill-fated from the first. Poorly equipped to start with, Brown's canyon expedition ended in disaster. President Brown himself was drowned in Marble Canyon, and a little later two others, Peter Hansbrough and Henry Richards, lost their lives.

Stanton, who had succeeded to the leadership of the expedition upon the death of President Brown, was then forced to abandon the project, but did so only with the resolve to come back as soon as possible and complete the task he had undertaken. It was not an easy matter to restore the morale of the railroad company's directors, but Stanton succeeded and immediately set to work arranging for the second expedition. Returning to the river some months later with an expedition much better organized than the first and equipped with superior boats of his own designing, Stanton successfully concluded the survey and established the feasibility of President Brown's proposal. Accordingly, he reported to the directors of his company that it was practicable to construct a railroad along the Colorado to the Gulf of California, and operate it with hydro-electric power supplied by the river.

But the canyon railroad which Stanton championed so ardently was not destined to be. With oil and electric power being made available to the Southwest, the need for coal was no longer particularly urgent. Economically, the financiers decided, the projected railroad

PREFACE xix

through the canyon was not to be considered. So ended the great project, but not Stanton's faith in its practicability, nor his keen interest in everything that was connected in any way with the romantic Rio Colorado.

After the collapse of the great plan for the canyon railroad Stanton became engaged in other engineering projects which kept him in active touch with the Colorado until the close of 1901. Some years later he completed the preparation of a comprehensive two-volume work which he called *The Colorado River of the West, and the Exploration, Navigation, and Survey of its Canyons, from the Standpoint of an Engineer.*

In addition to treating very fully the controversial matters with which this book is concerned, Stanton's manuscript gave his theory of the geologic history of the canyon country. It also dealt with all the various Colorado River expeditions and voyages, both Spanish and American. Naturally enough, it included, too, a detailed account of the Brown-Stanton river expeditions of 1889 and 1890. It was this very fullness and completeness which was to prove a stumbling-block in the way of its being accepted for publication. In after years Stanton himself came to realize the unwieldy length of his manuscript, as is apparent from the following, written by him in 1920:

"Such a work, written from the standpoint of an engineer, with its necessary arguments and discussions, I was well aware, had one great defect. After the manuscript was completed, in December 1909, it was submitted to several publishers. As I fully expected, each one returned it, with the explanation that he could not undertake its publication, as there would be no financial profit in it. On this account the manuscript has slumbered peacefully for years."

For a considerable time after the manuscript was first offered unsuccessfully for publication Stanton continued to gather material to add to his book. He was remarkably tenacious and untiring in his efforts to run down all possible information and proof on disputed points in the history of the river's exploration, particularly anything having to do with the James White and Separation Rapid affairs.

Few things in this world are more difficult than for a writer, having written a thing earnestly and to the best of his ability, to have to undertake a drastic revision of his own work. So it was, evidently, with Stanton. He had prepared his manuscript with infinite pains. He had written it from the standpoint of an engineer, and all the fullness of detail, the arguments and discussions which served to make the book of such unusual bulk as to scare out possible publishers seemed to him necessary to a logical and complete presentation of his subject. He could not bring himself to cut it down, and so went on hoping against hope that somehow the book might find a publisher who would bring it out at its full length.

When Robert Brewster Stanton died at his home in New Canaan, Connecticut, on Feb. 23, 1922, his

PREFACE xxi

book was still in manuscript form and without a publisher. Since his death it has seemed on several occasions that the entire manuscript might be printed, but each time some obstacle, usually the size of the book, has stood in the way.

Because these phases of Colorado River history, never settled satisfactorily elsewhere, are disposed of with reasonable conclusiveness—as far as they are ever likely to be—in Stanton's manuscript, it has been determined to publish only those portions of his work which bear upon the controversies in question. That, then, is in brief the *raison d'être* of this book: first, to present the results of Stanton's investigation of the James White story and so establish the validity of Major Powell's claim to being the first navigator of the Colorado: and, second, at the same time to refute the Major's charges of desertion against the three members of the first expedition who left him at Separation Rapid.

As for the James White story, curiously enough, this wildest of all Colorado River yarns clearly results from the over-zealous chronicling of a canyon adventure originally told quite simply by an ignorant, unimaginative, but honest man. On September 8, 1867, two years before Major Powell's first voyage down the Colorado, one James White was taken off a raft at the little Mormon settlement of Callville, Nevada, in a greatly exhausted condition. He had been prospecting, he said, with Captain Baker and George

Strole in the San Juan Mountains. Baker had been slain in an Indian attack. White and Strole had beaten off the attack and escaped to the river where that night they hastily constructed a raft of four eight-inch cottonwood logs, ten feet long, lashed together with lariats, and on it began to drift through the canyons. On the fourth day Strole was washed off the raft and drowned, whereupon White took the precaution of tying around his waist one end of a fifty-foot rope which he had attached to the raft. Ten days later he drifted to Callville and succor.

White knew that he had been through a very trying ordeal, and that he had somehow survived a perilous voyage. But just how remarkable that feat really was he did not appreciate until people began to impress upon him the sensational character of his exploit. It was claimed that he must have launched his raft somewhere on the upper Colorado, near the mouth of the San Juan and drifted on it through Glen Canyon, Marble Canyon, and Grand Canyon to Callville. His adventure so fired the imagination of some journalists of the day that some truly remarkable effusions were forthcoming, notable among which is the thriller by Major A. R. Calhoun, published in a volume entitled *Wonderful Adventures.*

Yellow journalism such as this, however, did more to discredit than to establish the claim that James White had actually traversed the canyons of the Colorado on an improvised raft. It has rested chiefly, through decade after decade of controversy, upon the story of Dr. C. C. Parry, a botanist attached to the Union Pacific survey party, who interviewed White at Hardyville several months after the alleged raft voyage through the canyons.

Dr. Parry's report, as submitted to his superior, General William J. Palmer, Chief of the Survey, and as subsequently published in the Transactions of the St. Louis Academy of Science, asserted that White had actually gone through the Grand Canyon. Dr. Parry's report shows two things clearly: that he thoroughly believed White had made a raft voyage of some 550 miles through the canyons, and that White himself (with sufficient encouragement from Dr. Parry) believed the same thing. It failed completely, however, to prove the alleged fact beyond serious doubts.

Yet so thoroughly was White convinced that he had made such a voyage that not all the argument in the world, not all the evidence adduced in later years by actual canyon explorers, could make him change his mind in the least. Forty years after his voyage he still maintained that he had made the entire trip *and had encountered but one bad rapid,* despite the fact that a number of subsequent expeditions offered incontrovertible evidence that there are not one but three hundred major rapids—any one of them capable of inspiring genuine terror in a man compelled to shoot it on a makeshift raft of three cottonwood logs.

Major Powell gave no credence to the James White story. Frederick S. Dellenbaugh, veteran of the second Powell expedition of 1872 and distinguished historian of the Colorado, dismisses the White claims in the following words:

"Just where he entered the river is of course impossible to decide, but that he never came through the Grand Canyon is as certain as anything can be. His story reveals an absolute ignorance of the river and its walls throughout its whole course he pretended to have traversed." Possibly it would be fairer to White had Dellenbaugh said "throughout the whole course over which Dr. Parry convinced him he had come."

In common with the others who have considered the raft yarn in the light of actual first-hand knowledge of the Colorado, Lewis R. Freeman, author of *The Colorado River* (1923) and *Down the Colorado* (1924), scoffs at the story of White's having travelled over five hundred miles on a raft in fourteen days. He would, Freeman estimates, have required several weeks to cover that distance, considering the extreme difficulty of keeping the raft in the current and retrieving it from rocks, sand bars, and eddies. Of White's assertion that he saved himself from the fate of George Strole only by tying himself to the raft, Freeman remarks, "I can think of no more certain preliminary to inevitable suicide than such an action— that thus bound he could have survived one major rapid of the Grand Canyon is hardly possible."

PREFACE

There is no complete record of the number of attempted Colorado River navigations which have ended in tragedy. The successful expeditions have been comparatively few in number since Powell pioneered in 1869. And it is all the more noteworthy that almost without exception the successful expeditions have been those which have been equipped with boats very specially constructed to withstand the rigors and perils of the river trip. Even so, occasionally there has been an instance wherein the boatman's skill has failed him momentarily, and these same steel or wooden boats have come to grief—dashed to pieces on the very rocks the fabled White raft drifted serenely by.

On the face of it, therefore, it would appear the most palpable nonsense, moonshine unadulterated, that four logs hastily lashed together with ropes could safely transport a man through hundreds of miles of treacherous waters in which well-equipped expeditions have not always been able to avert disaster. And yet the fact that Dr. Parry's report was published by the St. Louis Academy of Science carried such weight with one Thomas F. Dawson, Executive Clerk of the United States Senate, that he determined to see what could be done to gain for James White the recognition he felt was due him as the first navigator of the Colorado, and with somewhat amazing results. Through Dawson's efforts, on May 25, 1917, there was presented in the Senate of the United States (Senate Document No. 42, 65th Congress, 1st session) "An article giving the credit of first traversing the Grand Canyon of the Colorado to James White, a Colorado gold explorer, who it is claimed made the voyage two years previous to the expedition under the direction of Maj. J. W. Powell in 1869," by Thomas F. Dawson. And a resolution was passed declaring that to James White (not to Major Powell, to whom Congress later erected the monument in Grand Canyon National Park as the Colorado's first navigator) was due the recognition as the first conqueror of the perilous Colorado! Surely, as will more clearly appear later on in this volume, this must be accounted one of the drollest of all gestures of a Congress notoriously given to sporadic asininity.

For the moment, enough of James White. Let us consider his more distinguished contender to fame, Major Powell. First navigator of the Colorado was Major Powell, and *he never forgot it*. In fact, he developed an attitude of mind which led him to think that he had done all the exploring necessary. It became *his* river, and he bitterly resented the idea of anybody else attempting to duplicate his feats in having navigated the impetuous stream.

Describing his chief in the preface to *A Canyon Voyage* (1908), Dellenbaugh writes: "Major Powell was a man of prompt decision, with a cool, comprehensive, far-reaching mind. He was genial, kind, never despondent, always resourceful, masterful, determined to overcome every obstacle. To him alone belongs the credit for solving the problem of the great canyons. . . ."

Apparently John Wesley Powell was in rather thorough agreement with this last statement. Yet his first trip, in 1869, was made possible only because eight men shared with him the peril and the toil of exploring the unknown Colorado. And the expedition of 1872 comprised six men besides Powell. Yet for some reason or other Powell elected not to do full justice to the courageous fellows who went adventuring with him.

His government-printed report of 1875, *"Exploration of the Colorado River of the West,"* alludes but briefly to the second expedition, neglecting even to record the names of the men who accompanied him. For this strange circumstance Dellenbaugh apologizes somewhat ineffectually in his book, *The Romance of the Colorado River* (1906). Major Powell omitted naming the members of the second expedition, Dellenbaugh suggests, for the sake of preserving dramatic unity in the report as a whole.

Dellenbaugh further says that in 1902, the year of Powell's death, in some correspondence between the two, he informed the Major of his intention of writing the history of the second expedition, and that Powell then expressed the hope that the proposed history might put on record the names of the men who were members of that expedition—a belated mitigation of a state of affairs certainly doing no credit to the Major.

But even more unaccountable than Major Powell's

failure to record in his book the names of the men who accompanied him on the second voyage is a circumstance connected with the recording of the first expedition. The inscription on the Powell Monument, erected by Congress in Grand Canyon National Park in 1918 in commemoration of Major Powell's two trips down the Colorado, gives what purports to be the full roster of both expeditions. The bronze tablet reads as follows:

1869	1872
JOHN C. SUMNER	A. H. THOMPSON
WALTER H. POWELL	F. S. DELLENBAUGH
G. Y. BRADLEY	JOHN K. HILLERS
WILLIAM R. HAWKINS	STEPHEN V. JONES
ANDREW W. HALL	W. CLEMENT POWELL
	ANDREW J. HATTAN

ERECTED BY THE CONGRESS OF THE UNITED STATES TO MAJOR JOHN WESLEY POWELL, FIRST EXPLORER OF THE GRAND CANYON, WHO DESCENDED THE RIVER WITH HIS PARTY IN ROWBOATS, TRAVERSING THE GORGE BENEATH THIS POINT AUGUST 17, 1869 AND AGAIN SEPTEMBER, 1872.

It was but natural that this question should immediately arise: Why were there omitted the names of three men known to have accompanied the first expedition for about 960 miles, decidedly the greater part

PREFACE xxix

of the entire voyage, terminated at the mouth of the Virgin River? O. G. Howland, Seneca Howland, and William Dunn left the party at a point later named "Separation Rapid." Why should these three men, martyrs of the first attempt to explore the Colorado, be utterly ignored on a commemorative tablet set up by the United States Government?

This is a significant question, but perhaps the answer will be supplied if we seek first the answer to another: Why did the Howland brothers and Dunn leave the river? Major Powell gives his answer in his report. In his journal entry for August 27, 1869 Powell makes a point of the anxiety the increasingly perilous nature of the river is giving the party. Then, and with a fullness of detail strangely absent elsewhere, he tells of the abandonment of the expedition by O. G. Howland, Seneca Howland, and William Dunn. According to his account the three must stand as cowards and deserters.

During his life time nobody openly disputed Major Powell's explanation of the Separation Rapids unpleasantness. But investigations carried on by interested persons have since put the sincerity and truthfulness of that version in very serious doubt. True, the Major has not lacked supporters. Dellenbaugh has always staunchly defended his departed chief, and others have agreed with him. Lewis R. Freeman opens Chapter X of *The Colorado River* with this statement:

"The outstanding figure of the latest half-century of Colorado River history is that of Major J. W. Powell, and if the coming years bring any change in its status it will only be to leave it bulking bigger against the skyline of the future than of the present."

Farther on in the same chapter Freeman writes:

". . . it is not probable that much that is new will ever be brought forward either to authenticate or to discredit Powell's story of his historic journey. That the explorer did not put on record the personal dissensions that were rife in his party was due to two reasons: one, because things of that kind had no place in a government report, and the other, because he was too big a man to let the memory of petty passages due to the irritations of the voyage obscure his appreciation of the fact that it was to the devotion and courage of his men that he owed the triumphant outcome of the expedition. . . .

"The name of none of the four men * who left the party before the completion of the voyage appears on the Powell monument on the brink of the Grand Canyon, and rightly so. Possibly Powell himself, swayed by his heart rather than his head, would have ruled more leniently could he have been consulted. But in the calm, considered judgment of posterity the men were deserters, and as such did not qualify for commemoration on a monument to honour achievement and fidelity. The appreciation of the greatness of Powell's personal achievement will increase with the years."

* The fourth man to leave the party (the first in order of leaving, however) was Frank Goodman, who cut loose from the expedition at the mouth of the Uinta on July 5th.

Thus wrote Freeman in 1923, but there has been evidence in later years that even he has changed his mind very considerably since writing the foregoing paragraphs.

With the presentation of Stanton's findings in both of the matters at issue, it is earnestly hoped that in the minds of many others, confused by the clash of opinions advanced upon insufficient knowledge of the whole situation involved, these questions will be settled satisfactorily, once and for all time.

While, as has been said, this posthumous volume constitutes only a part of the exhaustive treatise that Stanton had hoped to publish, it seems fitting to set forth herewith his acknowledgment of his indebtedness to the many who had rendered him some service in connection with the preparation of his manuscript.

Over the date line of "New York, March, 1920," Mr. Stanton had written:

"It would require a volume to express my appreciation of the many courtesies that have been extended to me while gathering evidence on the historical part of this work, not only by friends and acquaintances but also by total strangers to whom I applied in this country, Mexico, and Europe. To all of these I here extend my sincere thanks. The assistance of the officers of the U. S. Geological Survey, from the Director down, has been most valuable and helpful. The Survey Bureau and in fact every other department of the Government have been most generous in supplying me with maps, photographs, official data, and assistance which could be obtained nowhere else.

"I am especially indebted to Mr. Wm. B. McKinley, E.M., for his research in old books and manuscripts in the original Spanish."

To this it seems necessary for the editor only to add his own word of thanks to Mr. Stanton's daughter and son-in-law, Professor and Mrs. Lewis S. Burchard, of New York City, not only for making the Stanton manuscript available for the present purpose but for their fine spirit of courtesy and cooperation at all times during the preparation of this volume.

JAMES M. CHALFANT

Columbus, Ohio,
May, 1932.

BY WAY OF EXPLANATION

So very difficult a matter it is to trace and find out the truth of anything by history.
—PLUTARCH

WHY another book on the Colorado River? It is true that stories of discovery and exploration as well as some of the adventures upon what is perhaps the wildest river in the world have several times been written. But of the true history of its exploration the half has never been told. Furthermore, some things have been written that had better been left unsaid. Of no other portion of our country, I might almost say of the world, is there so little accurately known, except by the few, as of this mysterious far Southwest and of the great river which flows through it, or possibly to state it more correctly, of which more inaccuracies have been told, written, and published.

My connection with the Colorado River of the West and its canyons has been purely that of an engineer in the discharge of his ordinary, everyday duties. I do not claim the glory of discovery or the heroism of exploration. Still, my trip by boat from Green River, Utah, through some of the canyons of the Green, all the canyons of the Colorado, and on down to tide-water at its mouth on the Gulf of California, for the purposes of a railway survey in 1889 and '90, together with the subsequent years, extending over 1901, which I spent upon the river and the after years spent in painstaking research, gives me, I think, the right to speak with some authority.

The manuscript for this book was first written at leisure intervals during the years from 1906 to 1909 while I was engaged in professional work in New York and California. As early as 1890 I had been asked by a publishing house to put into book form my experiences on the Colorado River of the West, but I never attempted to do so until 1906. I had published two magazine articles on our journey through the canyons, one in *Scribner's* of November, 1890, and the other in the *Cosmopolitan* of August, 1893, besides quite a number of newspaper letters, and a paper on the engineering features of our expedition of 1889 and '90, in the Transactions of the American Society of Civil Engineers, April, 1892. But it did not seem to me that a formal book was required to record the experiences of an ordinary railway survey, even though the nature of the route made it a perilous undertaking.

On the other hand, from the very first I had been considerably exercised over the many absurd and inaccurate descriptions of the Colorado River and its canyons and the accounts of their exploration that had been published, and had expected some day to put my notes, with other data which I had gathered, in a permanent form.

With this object in view, and with a desire for ac-

BY WAY OF EXPLANATION

curacy as far as it was attainable, I was led into a careful search through the records relating to the river and the canyon. This proved to be a much more extended undertaking than I had anticipated, resulting in a plan and scope for the hoped-for volume very different from the mere writing of an entertaining story of the adventures of myself and my companions on the Colorado in 1889 and '90.

As the research work went on, I was more and more astonished to find how curiously inaccurate in many if not in all instances were the accounts which early explorers had given to the public, however carefully and correctly their original field notes may have been kept. I was somewhat puzzled to account for this fact. If the explanation which I favor, namely, that the very air of the canyons distorted their perceptions and warped their truthfulness, should seem far fetched, I can only say that I am willing to let the reader draw his own conclusions.

The story of the Colorado River of the West and the history of the Southwest land through which it flows have from time immemorial been veiled in romantic mystery. The earliest accounts of this land and the canyons of this wild, raging river were given to the world by a people that produced the necessity for Cervantes' *Don Quixote*. It is perhaps natural that these Spanish records of achievement and discovery should read more like pure romancing than historical accounts of travel, exploration, and adventure.

BY WAY OF EXPLANATION

This is not surprising when one considers not only the character of the earliest discoverers and even of some of the later explorers but the nature of the country itself, full of strange forms and changing colors which inspire awe, wonder, and amazement, a country in which at every turn one is startled by scenes of grandeur beyond compare, of beauty beyond conception, of coloring so bold, so wild, and yet so delicate and harmonious as to turn an artist's brain with despair of being able to catch the mystic, hazy tints of opal, topaz, mauve, pink, and lilac that hang over its deserts, its mountains, and its canyoned gorges.

What is surprising is the fact that for nearly four centuries this same spell, leading, it would seem, to misrepresentation and want of candor, has influenced almost everything that has been told of or written about that fascinating land, particularly the stories of the exploration, navigation, and survey of the great River and its Canyons.

In the year 1530 the Indian slave Tejo fired the imagination and cupidity of his master, Nuno do Guzman, President of the Audiencia of New Spain, with his accounts of the riches in gold and silver of the "Seven Cities of Cibola." And in 1536, when Estevanico, the Barbary negro, crossed the whole country from the coast of Texas to eastern Arizona his marvelous stories led to that most romantic and heroic exploit, the "March of Coronado." Ever since those days there has been "a nigger in the woodpile" some-

BY WAY OF EXPLANATION

where in nearly all the tales that have been told of this wonderful land.

Of the river and the canyons it is the same, for ever since the year 1540, when the two boats manned by Nicolas Zamorano and Domingo del Castillo with their oarsmen piloted into the mouth of the Colorado River the three little ships under the command of Hernando de Alarcon carrying in their diminutive hulls the first party of white men, so far as any records show, to float upon the surface of its raging waters (raging even then and there with the fierce tidal "bore"), there would seem to have been a mysterious want of truth in almost everything that has been written about the navigation of that dangerous stream— dangerous in more ways than one.

And from the time, a little later in the same year, when Don Garcia Lopez de Cardenas and his little band of thirteen Spaniards stood upon the brink of the Grand Canyon and looked for the first time into the depths of that abyss, the mysteries and deceptions have developed and expanded. If we can trust their historians, and if I may be pardoned for using a bit of slang, that little band was so affected by the sight of the great chasm that they "went up in the air." And so it has ever been with nearly all who have looked upon these scenes.

This indulging in exaggeration has gone on for now nearly four hundred years, from the time of Friar Marcos de Niza to the present day and John Hance—

John Hance, that prince of canyon story tellers who lived so long at the Grand Canyon and for whom something can be said that can be said for few of the others: his stories interested and amused the tenderfoot, and they harmed no one, for there was not an unkind word in any of them. This almost unaccountable romancing is most noticeable in the writings of the early explorers of the river and its canyons. As Van Dyke has so truthfully said, "This is the land of illusions and thin air. The vision is so cleared at times that the truth itself is deceptive."

Indeed, it is in the very air one breathes. That peculiar veiling of pink, lilac, and pale yellow seen high up in the Grand Canyon seems to produce such an effect upon men's imaginations that they pour forth nothing but what is rose-colored, whether true or otherwise. Or if not this, then the dark blue shadows which dwell low down among the Archean rocks so affect their sense of truth and justice that dark deeds are either hidden and covered up, or are told or written in their blackest hue.

What is the real reason why from the sixteenth to the twentieth century explorers, scientists, university presidents, engineers, doctors, and so on down to travelers, trappers, hunters, and miners have told such wonderful stories, such doubtful tales, and such positive untruths of this land and this one wildest river in all the world?

As a partial answer I cannot refrain from again

BY WAY OF EXPLANATION

quoting from that most charming little book, *The Desert,* by John C. Van Dyke, in which it is remarked that as from the first so now "everyone rides here with the feeling that he is the first one that ever broke into this unknown land, that he is the original discoverer; and that this new world belongs to him by right of original exploration and conquest."

With this sense of proprietorship comes another feeling. Everyone has seemed to assume that after his journey no one else would ever have the temerity or even the right to set foot in the land or on the river, and that therefore what was told by him could never be disputed.*

---

* "And it being doubtful whether any party *will ever again* pursue the same line of travel," Lieut. J. C. Ives, in letter transmitting his report to the War Department, dated May 1, 1860.

"Ours has been the *first, and will doubtless be the last,* party of whites to visit this profitless locality. It seems intended by nature that the Colorado River, along the greater portion of its lonely and majestic way, shall be forever unvisited and undisturbed." Lieut. J. C. Ives's Report, 1857 and 1858, p. 110. (Ives forgot Cardenas, the many Spanish Padres, and the Patties, who went all over that section, some of them hundreds of years before 1858.)

"And James White, as the pioneer of this enterprise, will probably retain the honor of being the only man who has traversed through its whole course, the great Canyon of the Colorado." Dr. C. C. Parry, in his letter to Gen. Wm. J. Palmer, dated January 6th, 1868.

"The Grand Canyon will probably remain unvisited for many years again, as it has nothing to recommend it but its general desolation, or as a study for the Geologist." Closing comment (after going through the canyon) in Jack Sumner's original diary, written in 1869.

"The exploration of the Colorado River *may now be considered complete.*" Lieut. Geo. M. Wheeler, U. S. Geological Survey West of 100th Meridian, 1871.

This feeling that no one would do the same thing again is perhaps natural, and has been expressed to me by a number of men, members of the Powell and later expeditions upon the river, and I confess to having experienced it myself when in the Grand Canyon. But its effect upon me was soon swept away by a hope that I then entertained that some day thousands would be following me by means of the railroad for which I was then running a survey.

There has also been given another explanation. My old friend, the late Col. Charles D. Poston, of Phoenix, Arizona, writing years ago in *Kate Field's Washington,* put it in this way: "I am not so sure whether any one who has wandered through the northern part of Arizona and looked upon the wonders that nature has wrought—its gorges, its canyons, its mountains and its painted rocks, and upon its ancient stone cities, and the cliff dwellings of its canyons, is ever afterwards *quite sane.*"

It is of course not within the scope of this work to give a detailed history of the discovery of the river and of every man from that time to this who has been directly or indirectly connected with its exploration, its navigation, or its survey. As I trust I have already made plain, it is my intention rather to clear up and correct some of the untruths, the misapprehensions, and the misunderstandings concerning the character of the river and its navigation.

Something happened in the year 1867 on the Lower Colorado, at Callville, Nevada, which gave rise to a most remarkable story of what was supposed to be the first navigation of the Colorado River, through the whole length of its canyons, by one James White on a raft. Five years later there occurred an incident which is a peculiarly suggestive commentary on the James White legend. In the Spring of 1872 some trappers broke into Major Powell's stores cached at the mouth of the Paria. Carrying off oars and other things, they fixed up a raft and started down Marble Canyon. This little band—who and how many they were I do not know—succeeded in reaching the first bad rapid at Badger Creek, or perhaps they got to the next, a worse one, at Soap Creek, but at one or the other they were wrecked, barely escaping with their lives.

Having made some original and extensive investigation as to the truth of the whole White story, I will consider it somewhat in detail. My excuse for going so at length into the story of James White is that the account of his adventure has been in print many times. It has been published as a fact in a number of Government reports. It has also been printed in school text books. Furthermore, it was the subject of a document published by the United States Senate in 1917. As a matter of fact, it is my decided opinion that the story is entirely unfounded, as I hope to prove in the pages that follow.

It has been my fortune in years past to gather considerable original data of the early explorations of the Colorado River from some of the explorers themselves. In this connection I counted Friday the 13th of December, 1891, a lucky day for me, for as our party passed down the river that afternoon we were hailed by a man at one of the placer mines asking us for tobacco. He was entirely out, and only those who know what this means in a wilderness can appreciate his position. We landed a little below, gave him a pound of good tobacco, and discovered that the man was Jack Sumner. I had a half hour's talk with him, the beginning of a friendship that lasted to the time of his death. He gave us great good cheer, and the simple advice, "Go slow and be careful, and you will be all right."

I asked him many questions about the river below, particularly about the cataract where the three men left the party of '69, and going out, were killed by the Indians. I told him that was the one rapid and the one place on the whole river, which, judging from Major Powell's account,* I feared. He assured me there were no insurmountable difficulties at that point, and encouraged us in every way.

"But, Jack," I said, "the Major gives a long, detailed account of his experiences at that rapid," and read it to him from a typewritten copy I had with me. Sumner turned away with an air of resentment, remarking, "There's lots in that book besides the truth!"

I took Sumner's photograph and bade him goodbye, but I thought much of his remark as we followed down the canyons. When we finally reached the Howlands' and Dunn's rapid, a new light began to dawn on me; but I did not understand it fully until I met Professor A. H. Thompson,† in Los Angeles, in October, 1893.

I wish to make clear my desire to give full credit not only to my own faithful companions and assistants of 1889 and '90, but to all who have been connected with the heroic and perilous work of opening to the world a

---

* Major Powell's official Report, published in 1875.
† Prof. Thompson, Major Powell's brother-in-law, was with Powell on the 1871–72 expedition.

knowledge of the Colorado and the Canyon, particularly to the first explorers of the Marble and the Grand Canyons, Major John Wesley Powell and the brave little band that accompanied him in 1869. If, however, as an engineer telling a few very plain truths I shall seem rudely to brush away some of the dusty cobwebs that have gathered around the record of the past, I hope I may be pardoned, for the reader must remember that it is the engineer's wont to measure everything with transit, level, and chain.

And if in so doing I shall pluck and cast to the winds the feathers from the caps of some whose right to wear them has been long undisputed, the blame must be laid only to a desire to be accurate and just to all concerned.

CONTENTS

	PAGE
FOREWORD	vii
PREFACE	xv
BY WAY OF EXPLANATION	xxxiii

PART ONE
JAMES WHITE'S RAFT JOURNEY OF 1867

I	THE BASIS OF A LEGEND	3
II	DR. PARRY, LEGEND MAKER	16
III	I INTERVIEW THE HERO OF THE LEGEND	36
IV	RAFT JOURNEYS—IMAGINARY AND REAL	70

PART TWO
THE AFFAIR AT SEPARATION RAPIDS

I	MAJOR POWELL AS HISTORIAN	97
II	WILLIAM HAWKINS' STORY	138
III	JACK SUMNER'S ACCOUNT	164
IV	MAJOR POWELL'S VERSION REFUTED	214

ILLUSTRATIONS

Robert Brewster Stanton . . . . *Frontispiece*	
	FACING PAGE
Hero of a legend	8
Facsimile of the original James White letter of 1867 . . .	12
Below the junction of the Grand and the Green	24
In Glen Canyon	25
A stretch of the river below the Grand Canyon	32
In Marble Canyon	40
A rapid in Cataract Canyon	41
James White loses his companion in Marble Canyon . .	48
In Boulder Canyon	49
The wreck at Disaster Falls	56
Disaster Falls as the camera reveals it	57
In the Grand Canyon below Kanab Wash	64
Lava "Falls"	*80
Map showing the actual journey of James White . . .	81
Major John Wesley Powell	104
Almon Harris Thompson	112
William R. Hawkins	144
Rough water at Granite Ledge Rapid	160
John C. ("Jack") Sumner	176
Life preserver used by Major Powell on the first expedition	224

xlvii

PART I

JAMES WHITE'S RAFT JOURNEY OF 1867

CHAPTER I

THE BASIS OF A LEGEND

ON September 8, 1867, some men waded out into the Colorado River at the town of Callville, about one hundred miles below the Grand Canyon, and pulled to shore a rude raft bearing an emaciated, half-demented fellow who gave his name as James White. He later told his rescuers a tale of harrowing experiences following an Indian attack which compelled him to take to the river as a last desperate measure to save his life.

What is undoubtedly the first published account of White's journey is contained in a letter written by one E. B. Grandin from the mining town of El Dorado, just below Black Canyon, on September 9th, 1867, and appearing in *The Daily Alta California,* of San Francisco, on September 24th. Grandin, having obtained his information from the captain of a barge that had come from Callville on the 8th, records the story in this manner:

A man by the name of White arrived at Callville on the 7th instant, who has come all the way from Green River on a raft. He was badly bruised, nearly starved, and almost entirely naked. Judging from his appearance he has had a rough time, and according to his statements has had many hairbreadth escapes. He gives the following account: He (White) was in company with two men, who were formerly residents of St. Louis, Missouri; one of them was known as Captain Baker and the other was named George Strode [Strole]. They were prospecting together on a branch of the Colorado that they called San Juan River. It is between the Little Colorado and the Green Rivers. I think it is sometimes called "The Blue." About the 24th of last month they were attacked by Indians. Captain Baker was killed at the first fire, White and Strode got away, and succeeded in gathering some rope and some ten pounds of flour, and with their guns made for the river, At the river they were fortunate enough to find some drift wood, with which they made a raft, and embarked, preferring to trust to the river rather than to stay there and lose their scalps. Some three days after starting Strode was washed overboard and lost. White continued on alone and after running fourteen days reached Callville. Soon after starting the flour was either washed overboard or spoiled by getting wet, and he was seven days at one time with nothing to eat. Then he luckily struck some Indians, from whom he bought a dog, giving the Indians his revolver. He managed to make out on dog meat until he reached Callville.

He describes the Big Canyon of the Colorado as terrific, a succession of rapids and falls. Some of the falls, he thinks, are fully ten feet perpendicular. His raft would plunge over such places, rolling over and over, and he was compelled to lash himself fast to keep from being washed away from it altogether. He says there are rocky cliffs overhanging the river that he believes to be a mile and a half high.

THE BASIS OF A LEGEND

White thinks they were in the vicinity of what will prove to be good mines there, on the San Juan River, judging from the prospects they obtained.

Thus began the saga of James White, gold prospector. Those who because of their own personal encounters with the mighty river consider such a trip as that attributed to James White a sheer physical impossibility have, in most cases, given the claims for him but little serious attention. They have, however, attempted from time to time to account for his amazing story. Two theories have always been favored: first, that James White spun his romantic yarn simply for his own glorification; and second, that, having killed his two companions in a quarrel, he told his story upon coming among white men again in order to cover up his crime.

And yet, fifty years later, on May 25, 1917, to be exact, there was presented to the Senate of the United States and later printed as a Senate document, with illustrations, a paper representing the culmination in the growth of the legend centering about what can be demonstrated to have been, after all, a comparatively simple exploit. The title of this remarkable monograph, by Thomas F. Dawson, executive clerk of the Senate, leaves one in no doubt as to why it was written:

"THE GRAND CANYON,

"*An article giving the credit of first traversing the Grand Canyon of the Colorado, to James White, a*

Colorado Gold prospector, who it is claimed made the voyage two years previous to the expedition under the direction of Maj. J. W. Powell, in 1869."

The Dawson pamphlet runs to sixty-seven pages, containing more documents, dogmatic assertions, and expressions of personal opinion than had ever before been assembled for the purpose of upholding the raft story. Because of this documentation it is sufficiently plausible to convince those relatively uninformed concerning the whole subject of the Colorado River that James White, not Major Powell, deserves the honor of first going through the great canyons of the Colorado.

Furthermore, the circumstances of its publication are such that the average person would naturally believe Dawson's conclusions to have had the sober, considered approval of the government. Were this assumption a fact, one would find it somewhat difficult to reconcile with Congress' sanction of the Dawson pamphlet another action of that same august body.

For on May 20, 1918, a year almost to the day after Senator John F. Shafroth of Colorado presented the Dawson pamphlet to the Senate, there was dedicated in Grand Canyon National Park a monument on which is inscribed:

ERECTED BY THE CONGRESS OF THE UNITED STATES
TO MAJOR JOHN WESLEY POWELL, FIRST
EXPLORER OF THE GRAND CANYON

THE BASIS OF A LEGEND 7

After summoning what he considers his authorities for the James White story, Dawson in a way sums up his whole case by remarking that the problem resolves itself into a question of probabilities. Since men have gone down the Colorado in boats, he asks, why not White's raft, particularly at a time of low water?

It is largely on this same question of probability that every one of the actual navigators of the Colorado has refused to believe in the White claims. As one of them, I had always been decidedly incredulous of the serious claim that anybody had succeeded in traveling on a raft more than five hundred miles on what is possibly the most dangerous river in the world. And as for the low stage of water having any effect in minimizing the peril, I knew only too well that there are on the river two hundred rapids which are actually many times more dangerous at low water than at high.

Long before Dawson (who originally became interested in the White story through reading an account in the files of the *Rocky Mountain News,* at Denver) had prepared and published his brochure championing James White as the Colorado's first conqueror, I had investigated the whole matter with the utmost thoroughness, with the result that I was thoroughly convinced that it was unfounded in fact. As early as 1892 I set forth my position in the matter: Major Powell's expedition of 1869 was undoubtedly the first, and I, having successfully concluded a survey of all the canyons of the river in 1889 and 1890, lay claim to the distinction of being second down the great river.

From the time I first heard the White story, as I have said, I was incredulous. But I was not satisfied to attempt to settle the whole thing on the basis of probability. Instead I set to work to get at the truth by finding all possible evidence on both sides of the matter. And I have succeeded in this enterprise far beyond my early expectations. One piece of good fortune in my research endeavors brought to hand the version of the story as originally set down in writing by James White himself.

Eighteen days after reaching Callville, White wrote a letter to a brother in Wisconsin giving an account of his recent ordeal. This letter, constituting the only largely uninfluenced version of his adventure White has ever given, was first published in *The Rocky Mountain News,* in February, 1869. The April 1907 issue of *Outing Magazine* carried the popular version of the White story, accompanied by a photographic reproduction of the letter of 1867.

I was inclined at the time to doubt the genuineness of the letter in *Outing,* but by inquiry and investigation I succeeded in obtaining a loan of the original letter from J. E. Parry, and after careful examination there was no longer left any doubt that the letter was *bona fide.* The letter follows:

(Photograph taken in 1909)

HERO OF A LEGEND

James White at the age of seventy

THE BASIS OF A LEGEND

NAVIGATION OF THE BIG CANYON:
A TERRIBLE VOYAGE

Callville September. 26. 1867

Dear Brother it has ben some time senCe i have heard frome you i got no anCe from the last letter that i roat to you for i left soon after i rote i Went prospeCted with Captin Baker and gorge strole in the San Won montin Wee found vry god prospeCk but noth that Wold pay then Wee stare Down the San Won river wee travel down a bout 200 miles then Wee Cross over on Caloreado and Camp We lad over one day Wee found out that Wee Cold not travel down the river and our horse Wass Sore fite and Wee had may up our mines to turene baCk When Wei Was attaCked by 15 or 20 utes indis they Kill Baker and gorge Strole and my self tok fore ropes off from our hourse and a ax ten pounds of flour and our gunns Wee had 15 millse to woak to Calarado Wee got to the river Jest at night Wee bilt a raft that night Wee had good Sailing fro three days and the Fore day gorge strole Was Wash off from the raft and down that left me alone i thought that it Wold be my time next i then pool off my boos and pands i then tide a rope to my wase I wend over falls from 10 to 15 feet hie my raft Wold tip over three and fore times a day the thurd day Wee loss our flour flour and fore seven days i had noth to eat to ralhhide nife Caber the 8. 9 days i got some musKit beens the 13 days a party of indis frendey they Wold not give me noth eat so i give my pistols for hine pards of a dog i ead one of for super and the other breakfast the 14 days i rive at Callville Whare i Was tak Care of by James ferry i was ten days With out pants or boos or hat i Was soon bornt so i Cold hadly Wolk the ingis tok 7 head horse from us Joosh i Can rite yu thalfe i under Went i see the hardes time that eny man ever did in the World but thank god that i got thrught saft i am Well a gin and i hope the few lines Will fine you all Well i sned my beCk respeCk to all Josh anCe this When you git it

DreCk you letter to Callville, Arizona Josh ass Tom to anCy that letter i rote him sevel yeas a goe

JAMES WHITE

Of this letter someone has said, "Its matter-of-fact ruggedness gives one a vivid idea of this brave and simple-minded prospector who underwent one of the most remarkable experiences that ever a man lived to tell about." And yet, for convenience' sake perhaps it would be well to render this rugged prose into more easily understood language:

NAVIGATION OF THE BIG CANYON:
A TERRIBLE VOYAGE

Callville, September 26, 1867

Dear Brother: It has been some time since I have heard from you. I got no answer from the last letter I wrote to you, for I left soon after I wrote. I went prospecting with Captain Baker in the San Juan mountains. We found good prospects, but nothing that would pay. Then we started down the San Juan River. We traveled down about two hundred miles. Then we crossed over to the Colorado and camped.

THE BASIS OF A LEGEND

We laid over one day and found that we could not travel down the river, and our horses were sore-footed, and we made up our minds to turn back when we were attacked by fifteen or twenty Ute Indians. They killed Baker, and George Strole and I took four ropes off our horses, and an ax, ten pounds of flour, and our guns. We had fifteen miles to walk to the Colorado and got to the river just at night. We built a raft that night. We got it built about 10 o'clock. We sailed all that night. We had good sailing for three days. The fourth day George Strole was washed off from the raft and drowned. That left me alone. I thought it would be my time next. I pulled off my boots and pants. I then tied a rope to my waist. I went over falls from ten to fifteen feet high. My raft would tip over three or four times a day. The third day we lost our flour and for seven days I had nothing to eat but two rawhide knife scabbards. The eighth day I got some mesquite beans. The thirteenth day I met a party of friendly Indians, but they would not give me anything to eat. So I gave them my pistols for the hind parts of a dog. I ate one of them for supper and the other for breakfast. The fourteenth day I arrived at Callville, where I was taken care of by James Ferry. I was ten days without pants or boots or hat. I was sunburnt so I could hardly walk. The Indians took seven head of horses from us. Josh, I can't write you half I underwent. I saw the hardest time that any man in the world ever had, but thank God I got through it safely. I am well again, and I hope these few lines will find you all well. I send my best respects to all. Josh, answer this when you get it.

Direct your letter to Callville, Arizona.

Josh, ask Tom to answer that letter I wrote him several years ago.

(signed)   JAMES WHITE

This letter, written when White was thirty years old, is undoubtedly genuine, and yet it is inconceivable that the same man who wrote the body of the letter also wrote the heading, unless someone else helped him. How, otherwise, could the person who misspelled the simplest words in the letter itself head it with the flamboyant title *Navigation of the Big Canyon: a Terrible Voyage,* and spell every word correctly?

I submitted the original to Mr. David N. Carvalho, one of the most noted handwriting experts in the country. After careful examination he decided that both heading and letter were written by the same person, and at or about the same time, but that the heading was written in a more studied, hesitant manner, probably after the body of the letter had been written and signed. Guided by this finding, I offer the following explanation.

During the eighteen days after his landing and before the letter was written, White told Jim Ferry of his experiences, and how he had floated down a big canyon, between walls hundreds of feet high. And it (Boulder Canyon) was a big canyon to him, as it would seem to any one who had never seen a greater, just as its almost exact counterpart, Black Canyon, was years before to scientific men like Lieutenant Ives and the artist Egloffstein, who showed it immensely

Navigation of the Big Cañon
A terrible voyage

Callville   September 26 1867

Dear Brother it has ben some time since i
have heard from you I get no anser from the
last letter that i sent to you for i left soon
after i rote i went prospected with Captin Baker in
the San Won montin wee found very good prospet but
noth that Wold pay then wee wore Down the San Won river
wee travell down a bout 200 miles then wee cross over
on Colorado and Camp wee lad over one day wee
found out that wee cold not travel down the river and
our hors was sow fite and wee had may up our mines to
turene back when wee was attacked by 15 or 20 indes thay
kill Baker and gorge Strole and my self took fore ropes
off our horse and a ox ten pounds of flour and our
gunns wee had 15 miles to wak to to Colarado wee got
to the river about night wee built a raft that
night wee got it bilt a bot ten o clock tha night
wee saile all that night wee had good sailing for
three days and the forth day gorge Strole was wash
of from the raft and drown that left me alone i thought
that it Wold be my turne next i then pool off my
boots and pands i thonde til o raft tive was then
over folls from 10 to 15 feet hie my raft Wold
tip over thrie and fore time a day the tharde day
wee loss our flour and fore seven days i
had to eat to rathhide nife babes the 8 day i got some
muskit beans the 13 days a party of indes frendey thay
Wold not give me noth eat so i give my pistols
for hine pards of a dog i ead one for super and the
other breakfest the 14 days i rive at Callville whare i was

(Courtesy of J. E. Parry)

Facsimile of the original James White letter of 1867

tak care of by James ferry i was ten days with out pants
or boots i was soon forn it so i cold hadly walk the
ingis tok y hed horse from us I aust i can tel you half
i never went i see the hardes time that eny man ever
did in the world but thank god that i got thrugh
saft i am well a gin and i hope the few lines will find
you all well i send my felt respeck to all
I ask andr this when you git it i Direck you letter
to Callville Arizona

Post ask Ton
Ta ouly that letter
that i rote him
sevel yeas a go

James White

To...
Oct 1/67

Washaw W White
...
Wis

THE BASIS OF A LEGEND 13

large in his marvelous pictures ten years before White's time. While White and his companions at Callville were talking it over among themselves, someone among them must have suggested that White had come through THE BIG CANYON. Furthermore, reading the letter, more thrilled than was the modest White with the greatness of the adventure, one of them probably urged the adding of the heading, and in all likelihood stood over White and told him how to spell those difficult words. Under these circumstances he wrote them in a more formal, heavy hand, laboriously spelled out.

It should be noted also that in addressing the envelope White uses an even more formal hand, but that in this personal matter, not having his spelling master standing over him, he spelled his own brother's name "Joshway."

Viewed in the light of the case which has been made out for him by Dr. Parry and others, this letter of James White is notable for its failure to mention certain things that one might have expected. For instance, he does not mention the Grand, the Green, or the passing of the mouths of the San Juan and the Little Colorado rivers. Nowhere in the body of the letter does he use the word "canyon," or so much as refer to the great and impressive height of the walls. But the most glaring omission of all is the one thing that would have been most natural for such a man to write in such a letter in order to impress upon his brother that, as he says, he had "seen the hardest time that any man ever had in the world," that is, if he had known it even approximately. I refer to the great distance (five hundred and fifty miles) he had supposedly traveled on this raft.

Aside from these negative points, some of the positive assertions of the letter are decidedly open to question. That the party traveled two hundred miles down the San Juan Valley it is obviously impossible to believe. Even if we admit that they may have thus traveled 150, 100 or only 50 miles, this would in no way justify the claim that it was fifty miles from the San Juan to the Grand, or that they traveled the intervening distance in two days.

White says the party was attacked by Ute Indians. It has been argued by another that the Utes were not to be found below the Grand Canyon, and therefore they could not have killed Baker in the canyon leading down to the Pierce's Ferry. Since in my interview with him in 1907 White confused the Utes and the Hualapais at a point still farther down the river near Callville, I very much doubt if White and Strole, after the death of Baker, knew or cared whether Utes, Sioux, Hualapais, or Pai-utes were after them, just so they got away safely.

In the body of the letter there is one important point, namely, that he was on the river fourteen days. How could White have required that length of time to drift from Pierce's Ferry to Callville, a distance of about

THE BASIS OF A LEGEND

sixty miles? Well, in the first place he did *not* travel fifteen hours a day, as Parry affirms. That much is made clear in White's statement in later years. But it is more important to keep in mind that he was not necessarily making definite progress all the time he was afloat.

The very nature of the river prevented him from going steadily down stream so as to make the speed and distance Parry assumed he made. Parry himself records that White was held for hours in an eddy at the mouth of the stream which he assured White was the Little Colorado. Sometimes, driven into a big eddy, he might be held for an entire day. And again he might be carried for a distance *upstream*. Here my theory is based on my own experience.

I recall how in the expeditions of 1889–90, even with our well-equipped boats, ably manned by two powerful oarsmen and a skilled steersman, on one occasion my boat was obliged to run one rapid three times before we could get through it. Twice we were forced by the current into the big eddy, and carried a quarter of a mile up stream. Entering the rapid again just before the fall, we succeeded in getting through only on the third attempt. But even so, was White actually on the river fourteen days? I think not. It is my conviction that at the time White wrote his letter he knew neither where he had been, how far he had come, or how long he had been on the river—only that he had been for a while on a raft, somewhere on the Colorado.

CHAPTER II

Dr. Parry, Legend Maker

IN 1867 and 1868 a party under Gen. Wm. J. Palmer was engaged in making surveys across the continent on the 35th and 32nd parallels for a route extending the Kansas Pacific Railway to the Pacific Ocean at San Francisco and San Diego. The first full account of James White's alleged raft journey was included in Gen Palmer's official report to the president of the railroad. Originally it was submitted in a letter to Gen. Palmer from Dr. C. C. Parry, who signed himself "Geologist of the Survey."

Dr. Parry's account as found in Gen. Palmer's report, published in 1869, is as follows:

GRAND CANYON OF THE COLORADO

Account of the Passage of the Great Canyon of the Colorado, from above the mouth of Green River to the head of Steamboat Navigation to Callville, in the months of August and September, 1867, by James White, now living at Callville.

To Gen'l. Wm. J. Palmer,
    Director of Surveys, Kansas Pacific Railway.
SIR:—
    The railroad survey now in progress under your direction has afforded many opportunities for ac-

DR. PARRY, LEGEND MAKER 17

quiring valuable additions to our geographical knowledge of the unexplored regions of the far west, from original sources not accessible to ordinary map compilers.

Mining prospectors within the last twenty years, more adventurous even than the noted trappers of the Rocky Mountains, have hardly left a mountain slope unvisited or a water-course unexamined over the wide expanse extending from the Mississippi River to the Pacific Ocean. Could the varied and adventurous experience of these mountain men be brought into an accessible form, we should know nearly as much of these western wilds as we now do of the old settled portions of our country.

Among the geographical problems remaining for the longest time unsolved was the actual character of the stupendous chasms or canyons through which the Colorado of the West cleaves its way from its snowy sources to its exit into the California Gulf. Within the last ten years public attention has been frequently directed to this subject, and various Government expeditions have imparted reliable information in reference to the upper and lower course of this remarkable river. Lieut. Ives, in 1857-8 made a satisfactory exploration of the navigable portions of the Colorado, extending from its mouth to the Big Canyon, and since then a regular line of light draft boats have been successfully traversing these inland waters. Still, the Great Canyon remained a myth; its actual length, the character of the stream, the nature of its banks, and the depth of its vertical walls were subjects for speculation, and afforded a fine field for exaggerated description, in which natural bridges, cavernous tunnels, and fearful cataracts naturally formed a prominent feature. Now, at last, we have a perfectly authentic account of the character of the Great Canyon of the Colorado, derived from the lips of a man who actually traversed its formidable depths, and who, fortunately for science, still lives to detail his trustworthy observations of this most remarkable voyage. Happening to fall in with this man during my recent stay of a few days at Hardyville, on the Colorado, I drew from him the following connected statement, in answer to direct questions carefully written at the time:

James White, now living at Callville, on the Colorado, formerly of Kenosha, Wisconsin, was induced to join a small prospecting party in search of gold washings, in the San Juan region, west of the Rocky Mountains. The original party was composed of four men, under the command of Capt. Baker. This small party left Fort Dodge on the 13th of last April, and after crossing the plains, completed their outfit for the San Juan country in Colorado City, leaving that place on the 20th of May.

Proceeding by the way of South Park and the Upper Arkansas, they crossed the Rocky Mountains, passing round the head waters of the Rio Grande, till they reached the Animas branch of the San Juan River. Here their prospecting for gold commenced, and being only partially successful, they continued still farther to the west, reaching the Dolores and Mancos branches. The latter stream was followed down to the main valley of the San Juan, when they crossed over to the left bank, and followed down the valley 200 miles. At this point the San Juan River enters a canyon, to avoid which they again crossed to the right bank, and struck across a mountain range for the Colorado. In a distance of fifty miles, over a very rugged country, they reached this latter stream, or rather its main eastern branch, Grand River, being still above the junction of Green River—the united waters of which two streams go to form the Colorado proper. At the point where they first struck the river, its steep rocky banks were inaccessible, and they accordingly followed up the stream in search of a place where water could be procured. At an estimated distance of twelve miles they came upon a side canyon, down which they succeeded in descending, with their animals, and procuring a supply of water. They camped at the bottom of this ravine on the night of the 23rd of August, and on the morning of the 24th started to ascend the right bank to the table land above. In making this ascent they were attacked by Indians, and Captain Baker being in advance was killed at the first fire. The two remaining men, James White and Henry [George] Strole, after ascertaining the fate of their comrade, fought their way back into the canyon and getting beyond the reach of the Indians, hastily unpacked their animals, securing their arms and a small stock of provisions, and proceeded on foot down the canyon to the banks of the Grand River. Here they constructed a raft of dry cottonwood, composed of three sticks ten feet in length and eight inches in diameter; these were securely tied together by lariat ropes, and having stowed away their provisions and arms, they embarked at midnight on their adventurous voyage. The following morning, being the 25th of August, they made a landing, repaired their raft by some additional pieces of light cedar, and continued on their course. The river here was about 200 yards wide, flowing regularly at a rate of two and a half to three miles per hour. At this estimate they reached the mouth of Green River, and entered the main Colorado 30 miles from the point of starting. Below the junction the stream narrows and is confined between perpendicular rocky walls, gradually increasing in elevation. At an estimated distance of 40 miles from the mouth of Green River they passed the mouth of the San Juan, both streams being here hemmed in by perpendicular walls. From this point the canyon was continuous, with only occasional breaks formed by small side canyons, equally inaccessible with the main chasm. Still they experienced no difficulty in continuing their voyage and were elated with the prospect of soon reaching the settlements on the Colorado below the Great Canyon.

On the 28th, being the fourth day of their journey, they encountered the first severe rapids, in passing one of which Henry Strole was washed off and sank in a whirlpool below. The small stock of provisions was also lost, and when White emerged from the foaming rapids he found himself alone, without any provisions, and with gloomy prospects ahead for accomplishing his adventurous journey. His course now led through the Big Canyon, and was a succession of rapids blocked up with masses of rock, over which his frail raft thumped and whirled so that he had to adopt the precaution of tying himself fast to the rocking timbers. In passing over one of the rapids his raft parted, and he was forced to hold on to the fragments

DR. PARRY, LEGEND MAKER 21

by main strength till he effected a landing below in a shallow eddy where he succeeded in procuring a scanty supply of mesquite bread, barely sufficient to sustain life till he reached Callville, on the 8th of September, fourteen days from the time of starting, during seven of which he had no food of any description.

When finally rescued he presented a pitiful sight; emaciated and haggard with abstinence, his bare feet and legs literally flayed from constant exposure to drenching from water, and the scorching rays of a vertical sun, and his reason almost gone. Being, however, of a naturally strong constitution, he soon recovered his usual health, and is now a stout, thick-set, hearty man. His narrative throughout bears all the evidence of entire reliability, and is sustained by collateral evidence, so that there is no room to doubt that he actually accomplished the journey in the manner and within the time mentioned by him.

CONCLUSIONS

The following may be summed up as some of the new facts to be derived from this remarkable voyage as additions to our previous geographical knowledge of the hydrography of the Colorado River:

1st. The actual location of the mouth of the San Juan, 40 miles below Green River Junction, and its entrance by a canyon, continuous with that of the Colorado, above and below the point of junction.

2nd. From the mouth of the San Juan, to the Colorado Chiquito, 3 days' travel in the swiftest portion of the current, allowing a rate of 4 miles per hour for

15 hours, or 60 miles per day, would give an estimated distance of 180 miles, including the most inaccessible portions of the great canyon.

3rd. From Colorado Chiquito to Callville, 10 days travel were expended. As this portion of the route was more open, and probably comprised long stretches of water, it would not be safe to allow a distance of more than 30 miles per day, or 300 miles for this interval. Thus the whole distance traveled would amount to 550 miles, or something over 500 miles from the Green River junction to the head of steamboat navigation on the Colorado.

4th. The absence of any distinct cataracts or perpendicular falls would seem to warrant the conclusion that in time of high water, by proper appliances in the form of boats, good resolute men, and provisions secured in water-proof bags, the same passage may be safely made, and the actual course of the river mapped out, and its peculiar geographical features properly examined.

5th. The construction of bridges, by a single span, would be rendered difficult of execution, on account of the usual flaring shape of the upper summit; possibly, however, points might be found where the high mesas come nearer together.

6th. The estimated average elevation of the canyon at 3000 feet is less than that given on the authority of Ives and Newbury, but may be nearer the actual truth as the result of more continuous observation.

7th. The width of the river at its narrowest points was estimated at 100 feet, and the line of high water mark, 30 to 40 feet above the average stage in August.

8th. The long continued uniformity of the geological formation, termed "white sandstone" (probably cretaceous), is remarkable, but under this term may have been comprised some of the lower stratified formations. The contrast in reaching the dark igneous rocks was so marked that it could not fail to be noticed.

9th. Any prospects for useful navigation up or down this canyon during the season of high water, or transportation of lumber from the upper pine regions could not be regarded as feasible, considering the long distance, and the inaccessible character of the river banks.

10th. No other satisfactory method of exploration, except along the course of the river, could be adopted to determine its actual course, and peculiarly natural features, and James White, as the pioneer of this enterprise, will probably retain the honor of being the only man who has traversed through its whole course the great canyon of the Colorado, and lived to recount his observation on this perilous trip.

Respectfully yours,
C. C. PARRY.
Geologist of the Survey.

Hardyville, Arizona, January 6, 1868.

Since it is upon Dr. Parry's account that the whole fantastic structure of the James White story chiefly rests, we cannot examine it too closely. Under what circumstances did White narrate his experiences to Dr. Parry, and in what frame of mind were they heard?

During the weeks of his convalescence at Callville, White undoubtedly related his adventures repeatedly to his benefactors, Jim Ferry, Tillman, and others, telling of the canyon he came through. Warming to the subject also as he discussed it with White, somebody certainly must have exclaimed, "Why, you must have come through the Big Canyon!" Naturally, as the discussion went on, the conception of what White's experience must have been did not diminish. On the contrary his friends, by the simple magic of oft-repeated suggestion, soon convinced themselves that James White had actually traversed the Grand Canyon. Somewhat awed by the idea, and proud to know the hero of such a feat, they were not backward about advancing their belief.

Later when he accompanied some of these men to Fort Mojave, White told the officers there his simple, straightforward tale. But we can rest assured that his friends added their conviction about his having come through the Big Canyon. Duly impressed by the magnitude of the exploit attributed to White, these officers passed on the story to Gen. Palmer, who was immediately interested. Unable at the time to stay in the vicinity long enough to get in touch with the adventurer, Gen. Palmer instructed one of his official aides, Dr. Parry, to remain and interview White. Dr. Parry, then, met White at the specific order of the chief of the survey, despite the assertion in his report, "Happening to fall in with this man during my recent stay of a few days at Hardyville, on the Colorado, I drew from him the following connected statement, in answer

(*Photograph by F. A. Nims, 1889*)

BELOW THE JUNCTION OF THE GRAND AND THE GREEN

According to White, it was impossible to walk along the river at the point below what he assumed to be the junction of these streams

(Photograph by F. A. Nims, 1889)

IN GLEN CANYON

Here are flaming red sandstone walls over 1000 feet high, but White insists that in this canyon for a distance of 150 miles he passed only white sandstone walls 300 to 400 feet high

to direct questions carefully written at the time."

It can safely be assumed that James White related to Dr. Parry the same simple story that he had told to others. How may we account for the development of Dr. Parry's story, attempting as it does to cover the geography, topography, and geology of what was then the most unknown section of the country? There can be but one explanation, as I see it. Even before he met him, Dr. Parry had decided that White had actually succeeded in traversing the Grand Canyon, that he had made a journey of approximately 550 miles. His sole purpose in questioning White seems to have been to obtain a confirmation of this preconceived belief.

Beginning at the upper end to identify the geography, he asked White if he passed the mouths of any rivers at the beginning of his journey, and how fast he was traveling. White replied that, floating at the rate of about two and a half to three miles per hour, he and Strole passed the mouth of a river the following morning. At this estimate they reached the mouth of the Green River and entered the main Colorado, 30 miles from the point of starting, we are asked to believe. This estimate of distance and the assumption that it was the Green River are Parry's own conclusions.

Again, Parry undoubtedly asked White if he had passed any other rivers. And from White's replies he concluded, for instance: "At an estimated distance of 40 miles from the Green River they passed the mouth of the San Juan," and goes on to add that this is the "actual location of the mouth of the San Juan."

Actually, the real distance from the Green to the San Juan is about 145 miles. According to Parry's and White's own statements, that was the most sluggish part of the current. Taking Parry's minimum of two and a half miles per hour and fifteen hours per day, it would have taken White's raft four days to make the distance. It follows, then, that if White had ever traveled it and had given Parry any true description of the journey as to time, and Parry had figured the distance in that way, he would have seen that the actual distance to the San Juan could not be forty miles. But that would have upset his preconceived opinion that White had actually made the whole journey.

Dr. Parry makes White pass the "Little Colorado" "on the fourth day in the evening." It is not likely that White insisted to Dr. Parry that the Little Colorado came in on the right-hand side of the Colorado as he went down the river. Parry probably did not question him on that point at all. But as White told him that he passed some river "on the fourth day in the evening," Parry himself assumed that it was the Little Colorado.

Parry again says White estimated "the average height of the canyon as 3,000 feet." Here Parry was a little doubtful, for he says in his conclusions that this is less than given on the authority of Ives and Newberry. He nevertheless concludes that, although Ives

DR. PARRY, LEGEND MAKER 27

and Newberry measured the height with instruments, White's estimate "may be nearer the truth, as the result of more continued observation"!

Dr. Parry then figures out how far White traveled, making it about 550 miles. He also undoubtedly questioned White on the geology of the canyon. White told him that the walls were "white sand rock." And Parry forthwith accepts this geological survey of 550 miles of canyons by an uneducated prospector as against the positive statements of the distinguished geologist, Dr. J. S. Newberry, and hastens to declare, "His narrative throughout bears all the evidence of entire reliability and is sustained by collateral evidence."

I have always been puzzled to understand why the "geologist" of Gen. Palmer's expedition should thus attempt to discredit the geological reports of Dr. Newberry, or question Ives's measurement of the height of the Canyon. What was Dr. Parry's duty, under all these circumstances? Plainly, it was to examine carefully all existing records and prove or disprove White's story by such evidence and testimony as was then accessible. Did he do it? If he had, what would he have found?

The evidence could not, perhaps, have been obtained at Fort Mojave or at Hardyville, but on his return to Washington and the East, before Gen. Palmer's report was published in 1869, and before the Academy of Science of St. Louis had been furnished the story,

Parry should have reread the report of Lieut. J. C. Ives of his explorations in 1857–58, published more than seven years before. (*Report upon the Colorado River of the West,* Washington Printing Office, 1861.) And he certainly ought to have read or reread Chapters five and six of the excellent report of Dr. J. S. Newberry, whose geology of the region has long been proved correct in all its essential features.

Did Dr. Parry examine Newberry's report for the purpose of verifying White's narrative as he related it? Charity for him compels me to say he did not. If he did, I leave others to draw their own conclusions. He knew of this work, as is seen in the fact that in his letter to Gen. Palmer he quotes from it twice in his prologue and once in his conclusions. The truth is that both Ives and Newberry had published positive refutation of the whole case which Dr. Parry was so earnestly making out for James White. But he, "untiring as a searcher after truth," never once looked into their reports to see if such collateral evidence sustained or disproved his story. The very fact that he did not constitutes the strongest sort of indictment against him.

The granite shown in Dr. Newberry's geological section of the Canyon is just about a hundred miles above where Parry places it, sixty miles above Callville. This fact should have set him to thinking again. The "igneous rock" in reality begins only some eighteen or twenty miles below the Little Colorado, where according to Parry, "the contrast is so marked that it

DR. PARRY, LEGEND MAKER 29

could not fail to be noticed." Hence White should have seen it on the fourth day of his journey, or at least on the fifth day, according to Parry's chronology, instead of on the "last two days of the Canyon."

Had Dr. Parry carefully and conscientiously reread Lieut. Ives's report, with Newberry's accurate geological descriptions, he would not have been led into the further ridiculous error of concluding that a half-starved, half-mad miner afloat on a raft could come "nearer the actual truth as to the height of the Canyon" as the result of "more continuous observation" than could Ives by instrumental measurement. If he had looked for it, he might have found Ives's statement: "The barometric observations upon the surface of the plateau and at the mouths of Diamond and Cataract rivers showed that the walls of this portion of the Canyon are over a mile high."

On just what statements by White himself is Dr. Parry's story founded? The original notes taken down at the time of the interview should throw some light on the question. A search extending over the better part of ten years was finally successful when in December 1916 I received them from Dr. Parry's nephew, Mr. John E. Parry, of Glens Falls, N. Y., the same source from which, six years previously, I had obtained a loan of White's letter to his brother Joshua. With my own interpolations in the interests of clearness in brackets, there follow the succinct notes of that historic interview at Hardyville, Arizona:

"James White [formerly of] Kenosha, Wisconsin. Started from Ft. Dodge, April 13, [1867] with a prospecting party under Capt. Baker. Came to Colorado City; left 20th of May for San Juan [country]. Struck Animas [and] Dolores [Rivers]. Mancos Canyon followed to San Juan [River]—down that [stream] 200 miles [then] crossed to north side. Crossed over Mts. to Colorado [River] 50 miles. Went up [Colorado] 12 miles to canyon. Went down to Colorado 12 [miles]. Henry Strole [the third member of the party]. Capt. Baker killed. Went back into canyon, unpacked, took 10 lbs. flour and coffee. 24th. Aug. fight [occurred]. Went to mouth [of] canyon [at Colorado River], built raft [of logs] 8 inch [thick], 10 ft. [long]—3 [logs] tied up with lariats. River wide and still, [with] small bottom [land] 25th, stopped and repaired raft; passed [mouth of] Green [River]—[in] 30 miles. After leaving [mouth of] Green [River]—Canyon [begins]. Traveled 40 miles to [mouth of] San Juan [River]. Laid up night [of 25th] 26th, traveled all night—40 miles. 27th [traveled] all night. 28th [the] 4th [day of trip] came to rapids and Strun [Strole] was washed off and drowned [at] 3, P. M. Lost provisions. Kept passing rapids—25 or 30 a day. Passed [mouth of] Colorado Chiquito 4th [day] in evening. Continuous rapids to 100 miles above Callville. Rock in Canyon white sandstone [For] 2 days in foot of Canyon [rock was] volcanic. Reach Callville 8th Sept. Line of high water mark 30 to 40 feet [upon wall]. Width of river in canyon—narrowest [part] 100 ft. Height [of walls of Canyon] 3000 ft. Rapids caused by fallen rocks. One fall 10 ft.? Many whirlpools and

DR. PARRY, LEGEND MAKER 31

eddies. Stopped in an eddy [at] mouth of C[Colorado] Chiq [Chiquito] 2 hours—prayed out. Shape of canyon perpendicular for several 100 ft. [up], then flares out. Course of river very crooked. Raft bumping on rocks. Same character or rock [white sandstone] through [all] the main canyon."

As for the "direct questions carefully written at the time," nobody, to my knowledge, has ever found them. I assume them to be non-existent, and my theory is this: Dr. Parry was somewhat careless in punctuation and sentence structure. He wrote in his report, "I drew from him the following connected statement, in answer to direct questions carefully written at the time." What he intended to say was, I believe, "In answer to direct questions I drew from him the following connected statement, carefully written at the time."

This being the case, is the sheet of notes which we have a statement "carefully written out?" My answer is two-fold. As regards the mechanics of writing, no. But as a set of skeleton notes of an interview, to be expanded and written out in full soon after, yes. That is, provided the record is correct as to what White related regarding his trip, for that is what it purports to be. And in the main it is undoubtedly correct. Such skeleton notes of an interview made at the time, are very common and perfectly proper. The sheet of notes under consideration is perfectly clear and understandable to me, just as it is written. Still, is it a correct record of what White related in answer to the direct questions? Let us look into this point for a few moments. To my mind, these original notes do not contain a single item in the form of proof of his story, as being accurate in detail or even correct in general, that is, as relating to the whole supposed journey of 550 miles.

Nowhere in the notes is it recorded that White mentioned the Grand River. It would seem clear that upon writing his report and quite possibly after consulting a map of the region Dr. Parry decided to put White on the Grand. This would enable him to float past the mouth of the Green, that being where Dr. Parry had decided White ought to have been. Small wonder that in after years White candidly admitted that he never knew he had passed the mouth of the Green until he was so informed by Dr. Parry.

The only distances traveled on the raft as recorded in the notes that by any possibility could have come directly from White are two aggregating seventy miles. One of these Dr. Parry says he himself calculated. No record even remotely relating to the supposed total distance of 550 miles is given in the notes. Considering the half-demented state in which White arrived at Callville, I am sure he could not have given Parry any idea of the whole distance he had traveled in his raft. Nor did he ever claim to know. It is true Dr. Parry does not claim in his report that White gave him this distance, but rather that he himself calculated it. But the astonishing thing is that there is not one par-

(Photograph by R. B. Stanton, 1890)

Stretch of the river about twenty-five miles below the Grand Canyon which is in all probability the place where White feared he was going to be swept into a tunnel

DR. PARRY, LEGEND MAKER 33

ticle of information in the notes on which to base such a calculation.

The statements in the notes relating to the geology of the canyon—the white sandstone and volcanic rocks; the high-water marks; the shape of the canyon; the width of the narrowest part; the character of the river and its many rapids and whirlpools; White's experiences on the river, his raft bumping on the rocks, his being caught in the whirlpool where he prayed himself out—all these are undoubtedly correctly reported. Moreover, they are entirely true of that part of the Colorado, as White later described it to me, where he actually went.

The notes record "One fall 10 ft.?"—in accordance with what White has always vigorously insisted, that is, that he encountered but one big rapid, or fall. Why the interrogation point? Did Parry doubt White's word that there was only one big rapid in the whole course of the 550 miles he was attempting to convince himself White had come? Or was he merely indicating his uncertainty as to the height of that one fall? My surmise is that White succeeded in convincing Dr. Parry that through all the canyons there were not any heavy rapids, save one. The good doctor, not believing White's statement as to the height of the one fall, deliberately omits all mention of it in his official report, remarking instead upon "the absence of any distinct cataracts or perpendicular falls."

With regard to the height of the canyon walls the

notes record "3,000 feet." I am satisfied that White told Dr. Parry just what he told me forty years later: "Where I went, the rock was all white and yellowish sandstone . . . and the walls were 300 or 400 feet high. The conclusion is unavoidable that Dr. Parry, knowing of Ives' and Newberry's measurements and convinced that White had come through the Grand Canyon, argued this point with him, and set down in his notes the estimate of 3,000 feet as a sort of compromise figure.

It is apparent that Dr. Parry did not give the whole of White's testimony to Gen. Palmer, but that instead while he was engaged in writing his report he deliberately doctored his notes in an effort to convince himself and others as to just where White's journey took him, what he ought to have seen, as well as what he ought not to have seen.

Primarily he saw in White good copy. He represented a journalist's find, and Parry determined to make a great story of him. I do not mean that he set to work to manufacture the story out of whole cloth—not at all. There was a basis of truth to begin upon, but Dr. Parry was determined that his effort should be of epic proportions. And it was.

With those original notes before me, carefully studied in the light of all other evidence advanced in all these years, I am more nearly certain now than I was in 1908—if such a thing could be possible—that Dr. C. C. Parry, in fact a doctor of medicine, officially designated as Gen. Palmer's botanist, and self-styled "geologist" of the survey, stretching a real journey of about sixty miles above Callville to an alleged journey all the way from the Grand River through the Grand Canyon, a distance of some 550 miles, was the father of the whole outlandish James White yarn so vigorously but misguidedly championed by Thomas F. Dawson and others to this very day.

CHAPTER III

I INTERVIEW THE HERO OF THE LEGEND

OVER a period of forty years the old story had been rehashed and reprinted many times. Not one word as to the truth or falsity of the original tale came from White, although he had been living in robust health all the time the controversy had been going on around him. The honor thrust upon White by others, particularly by Dr. Parry and Gen. Palmer, seems never to have turned his head.

One of the members of our Colorado River party in 1889, writing to me about White's letter, said that he had once met White some years before and talked with him, that he believed he was still living, and gave me his address. I immediately got into correspondence with a James White at Trinidad, Colorado and on my return from California in the fall of 1907, I stopped off there.

My object in visiting White was to get at the truth of his story, and, if he were able to demonstrate the fact, to give him the credit of being the first man to navigate the Colorado through all of its great canyons. On the other hand, if he failed to establish the correctness of his story, I was determined to see if in any way I could unravel the mystery of its telling.

THE HERO OF THE LEGEND 37

I reached Trinidad on the morning of September 23rd, and started out to call on Mr. White. Approaching the neighborhood of his home, I stopped a wagon with two men in it to inquire the way. Noticing on the wagon the sign "James White, Express," I asked the driver if his name was White, and he replied that it was.

I introduced myself, asked when I could see him and when I could have a talk in accordance with my recent letter to him.

"You want me to tell you about the canyon?" he asked.

"Yes, sir," I replied.

"Well, is there anything in it for me?"

"You mean any money?"

"Yes," White replied. "I won't tell about the canyon unless there is money in it for me. People come from everywhere to get me to tell them about the canyon. One newspaper man came from Denver, and one man came all the way from St. Louis, but they said there was nothing in it for me, so they went away without getting anything at all."

"Why, of course there is," I hastened to say, "I would not ask you to haul an express load for me without paying you for it, and I don't want to take your time telling me about the canyon without my paying you for that. When can we have a talk?"

"Well, I'd rather have it in the evening. If I go with you now I'll have to put my horses in the barn and lose my day's work. I'll be home at six o'clock."

I found the original raft voyager a vigorous man of seventy who but for his sparse gray hair would not have been taken for over fifty. That I had met the genuine and only James White I satisfied myself by a few questions at the beginning of my interview. Before this, however, I had employed as stenographer to take down the interview a young man, who, knowing the subject of my investigation, but not knowing the name of the man we were going to see, volunteered to call my attention to a certain James White, almost his next door neighbor, whose daughters had told him of their father's journey down the Colorado River in 1867. The man he referred to and the man I interviewed proved to be one and the same person.

That evening I called upon White. After certain preliminaries, including my giving him my check for $25, he cleared off a table for the stenographer, Roy Lappin, a fellow-townsman, and proceeded to dictate his story to him. In order to bring out certain points I interrupted occasionally. How far, if at all, my questions may have influenced the story let the reader judge for himself.

Certainly it is not often that after forty years of dispute and disagreement the chief actor in the episode in question (and he thought long since dead) can be found and a first-hand account of his exploit obtained. This is James White's story as I heard it from his own lips. I trust it may not be without interest and historical value.

WHITE: During the year 1867 [1866] and part of 1868 [1867] I was in the employ of the Barlow & Sanderson Stage Company, running from Kansas City to Santa Fé, New Mexico. My run was between Fort Dodge and Cimarron Crossing. My particular companions at this time were George Strole, Joe Goodfellow, and a Captain Baker, as he was generally known. Although we called him Jim Baker, his real first name was never known to any of us. Early in April 1868 [1867] we four decided to leave the Stage Company and go west on a prospecting trip. In order to supply ourselves with means of transportation, we decided to steal some horses from the Indians who were camped on Mulberry Creek about one hundred miles east of Cimarron Crossing.

We left Cimarron about the middle of April and went east about one hundred miles to Mulberry Creek where the Indians had a band of horses. That is, Captain Baker, George Strole, and I did. Joe Goodfellow started with us but only went half way and then turned back. He joined us again a few days later, on our trip west. We three, Baker, Strole, and I, hid in the willows along the stream all day until about nine o'clock that night, when the moon came up so we could see the horses. We started towards the camp, and although the dogs in the Indian Camp began to bark, we each succeeded in catching a horse and mounting. We rounded up the band and cut out some thirteen head and drove them as fast we we could gallop away up the river [the Arkansas] towards Colorado City, which was our destination.

The Indians from whom we had stolen the horses did not overtake us, but we learned afterwards that the next night they attacked the stage barns at Cimarron and drove off thirteen head of horses to replace the ones we had taken from them. Joe Goodfellow rejoined us. And when we had traveled two or three days Goodfellow wanted to divide the stock equally between the four men of the party. Although Baker was captain of the party, he said that he only had one say in the matter, but that he would put it to a vote. Baker asked me if I was willing to give up one third of the stock to Goodfellow, who had not been with us when we stole them, and I told him I was not. He then asked George if he would give up one third, and George said no. We argued the thing out, but could not agree with Goodfellow, but we all four went on together to Colorado City [just west of the present Colorado Springs, Colorado], where we completed our outfit.

We sold one mule for $200 and sold Goodfellow two horses and a colt for $250, bought grub and other necessary things, packed our animals and started on west, leaving Colorado City the latter part of May. We first went to Salt Springs and one

(*Photograph by F. A. Nims, 1889*)

IN MARBLE CANYON

Through this canyon James White professed to have gone, and yet to have seen instead of its red, blue, green, and yellow marble walls, 2000 to 3000 feet high, only "white sandstone," 300 to 400 feet high

(*Photograph by J. A. McCormick, 1891*)

A RAPID IN CATARACT CANYON

This is a sample of the eighteen mile stretch in Cataract Canyon through which White alleged he floated at the rate of two and a half miles an hour on smooth water

day's march beyond, where we camped. Baker had put me in charge of the pack animals. We had one mule in the outfit and we put two hundred pounds on the mule and packed the other animals equally. That night near Salt Springs I took 25 pounds of flour off Goodfellow's horse and baked most of it into bread. The flour not used I put back on to Goodfellow's horse, and was going to put the bread on the horse also, when Goodfellow said the bread ought to be put on the mule. I said it ought to go on the horse because I had taken the flour off the horse.

We quarreled over it. We drew our revolvers and passed five shots between us. I hit Goodfellow twice, shot him once in the leg and once in the arm. We disarmed him, and put him on his horse and I took him to a house. I told the woman to take care of him, that he had two or three hundred dollars and two horses, and that he would pay her when he got well. As we parted he said to me that when he saw me again he would kill me, and I told him when he saw me to go ahead and shoot. I bade him good-by and left him, and Baker, Strole, and I went on our journey. Our route was from Colorado City to Salt Springs, across the range to the Rio Grande by Wagon Wheel Gap to the head waters of the river, and then across the main range [continental divide] to Rickey [Eureka] Gulch. We stayed in this neighborhood about a month, prospecting.

STANTON: What date was this, and when did you

42 COLORADO RIVER CONTROVERSIES

leave?

WHITE: Just the exact date when we got there or when we left I cannot say, but I remember we were there in camp on the 4th of July. We worked on a little stream near where Silverton, Colorado is now located.

From Silverton we went west over the mountains and then traveled south and west until we struck the San Juan river, about the first of August. We came to the San Juan about twenty-five miles above its mouth.

STANTON: Are you sure it was twenty-five miles above the mouth where you first came to the San Juan?

WHITE: It may have been more. I cannot be exact, but think the place we struck the San Juan was twenty-five or thirty miles above its mouth. We crossed the San Juan, and continued our journey southwest from there until we struck the head of a canyon. We went down the canyon one day's march and camped.

STANTON: Wait a moment (drawing a sketch in my note book).

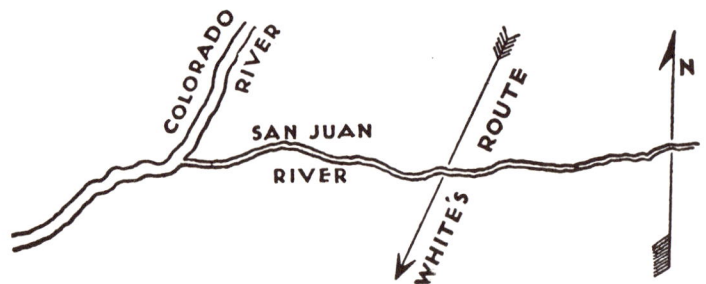

Here is the San Juan and here is the Colorado.
You traveled southwest or south and crossed the San
Juan and then continued southwest until you struck
the canyon which you went down?

WHITE: (Examining the sketch.) Yes, sir.

STANTON: How long were you going from Silverton
to the San Juan River?

WHITE: We were five days. We traveled three days,
two days from Silverton to a lake where we stopped
and fished two days, and then one day from the lake
to the San Juan.

STANTON: How far was it from the San Juan to the
camp in the canyon?

WHITE: I don't know how far it was, but it took us two
days [actually forty-five days] to get there, one day
from the San Juan to the head of the canyon, and one
day down the canyon to where we camped.

STANTON: How far did you usually travel in a day?

WHITE: Twenty, twenty-five, or sometimes thirty
miles in a day.

STANTON: Would twenty-five miles be a fair average
day's ride?

WHITE: Yes, I think it would. Baker went down the
Canyon and George went up looking for water.
When they came back I told them I saw an Indian.
They said I was crazy and only imagined I saw one.
The next morning we started back northeast, the
way we had come in. We had gotten half way out
when I heard a shot and Baker, being ahead,

stopped. I cried out and said, "What are you shooting at?" and George said, "The Indian!" I hollered to Baker and asked him if he was shot. He fell on his face and never answered. The Indian had shot him just above the heart.

I took Baker's horse and George and I went back down the canyon to our camp the same way we came up. We took our lariat ropes, overcoats and other things off the horses and left them there and went on down the canyon until we struck the Grand River.

STANTON: How far was it from the camp to the "Grand River"?

WHITE: The distance from our camp to the Grand River was about twelve or fifteen miles, and where we first struck it was about two miles above the mouth of the Green River. That night we built a raft of four dry cottonwood logs six or eight inches thick and eight to ten feet long, tied together with our lariat ropes. We set sail that night about ten or eleven o'clock. It was a moonlight night. We were on the raft two or three hours when we struck the mouth of the Green River.

STANTON: Did you travel every night as well as day when you were on the river?

WHITE: No, we traveled all that first night, but after that we tied up every night.

STANTON: In that distance were there any rapids in the Grand or Colorado Rivers?

WHITE: No, sir. We floated down the river for four

days on smooth water below the mouth of the Green River, without any trouble and without any rapids of any kind. The water was so smooth that we sat on the raft with our feet dangling in the water and just floated along.

STANTON: How far did you travel in that time?

WHITE: I don't know. We traveled the first night and the following four days at the rate of two or two and a half miles per hour. Near night of the fourth day we struck our first rapid.

STANTON: Was this rapid a large one?

WHITE: The rapid was not a large one. It was a small rapid, but our raft was so small that it tipped over, and I was thrown into the river. George caught me by the hair and saved me from drowning.

STANTON: Was this the first accident you had on the river?

WHITE: Yes, this was the first accident we had on the river. We dropped below the rapid and camped for the night. Here we found some cedar logs at the mouth of a side canyon which were one foot through and fifteen or sixteen feet long. We built a second raft which was eight feet wide, tied together with the lariat ropes we took off the first raft. This raft we built in the early morning of the fifth day. We set sail again on it about nine o'clock in the morning. We sailed about an hour, when we struck another rapid and George took a pike pole to fend the raft off the wall, and the raft went into a whirlpool and George was washed off. I saw him in the water and called to him to swim ashore, but he sank and I never saw him again. This happened about ten o'clock in the morning and from then I sailed on alone all that day.

On the sixth day, the day after George was drowned, the raft was drawn into a whirlpool and spun round and round. I paddled and paddled, but I could not get out. I was in that whirlpool for about two hours, and I thought I was lost. I prayed to God Almighty to help me, and immediately the raft was swung out of the whirlpool like a shot. This was the first prayer I had ever said in my life. I continued on down the river that day and the next, the seventh day I had been on the water, passing over many rapids, all of them small, and none of them great rapids. But on account of my raft being small and easily upset, it turned over in almost every rapid, so that when I climbed up I was first on the top of the raft and then on the bottom, which had turned up as the top.

On the eighth day my raft got stuck between two rocks and I got out on one of the rocks and pried out the raft, and it went on down stream and left me on the rock. I had a rope fifty feet long tied to the raft and tied around my waist. I had sailed this way, tied to the raft, with a fifty-foot rope ever since George was drowned. I took hold of the rope and jumped into the river and pulled myself to the raft and climbed on.

STANTON: What kind of rapid was that one?
WHITE: It was not a rapid, it was just water pouring over rocks. During the fifth, sixth, seventh, and eighth days there were not any heavy rapids, only small rapids and swift water.
STANTON: How fast did you travel over those small rapids and the swift water?
WHITE: I don't know. Faster than over the smooth water below the mouth of the Green River. Perhaps three or three and a half miles an hour. On the ninth day I went over a rapid and the raft broke apart and let me into the river. I held on and dropped below the rapid, pulled the broken raft out to shore, and tied it up again.

The tenth day I went over one big rapid with a fall of twenty feet, perhaps more. The raft turned over three times with me, and turning over and over wound the rope with which I was tied to it around the raft so tight that I had to untie it off my waist before I could climb on the raft. After I got over this rapid I went into camp.
STANTON: What time of day was that?
WHITE: That was about four o'clock in the afternoon. The next morning, which was the eleventh day, I got up early and set sail at day light. I traveled until about twelve o'clock. I came to an island with the water running on both sides, and my raft went on the weakest side where the water was shallow, and it stuck on the rocks and small boulders so I could not

get it off. I waded to the island through water sometimes over my ankles and sometimes up to my arm pits.

STANTON: How large was this island?

WHITE: It was about an acre. When I got on the island the sun was so hot, and not having any clothing except a coat and part of a shirt I was all blistered and sore. I got under a big log and stayed there in the shade for about three hours. I then went out to the raft and untied the ropes, and came back to the island again. I found four sticks of dry cottonwood on that island, about eight feet long and eight inches through, and I pulled them down to the river below the rapid and tied them together and made my third raft, and then went back under the tree on the island and slept until the next day.

The twelfth day I sailed all day on still and some swift water. My raft was so small that it tipped over several times, but the rapids were all very small and none of them heavy ones. On the thirteenth day I got up and the moon was so bright I thought it was day, but it was about three o'clock. I set sail and pretty soon I heard some voices and I hallowed and they hallowed back to me, and pretty soon four or five Indians waded out in the water to their waists and got my raft and pulled me in to shore. I asked them if they were friendly and they said yes. When I got to shore I saw about seventy-five Indians and asked them and they said they were all friendly. I

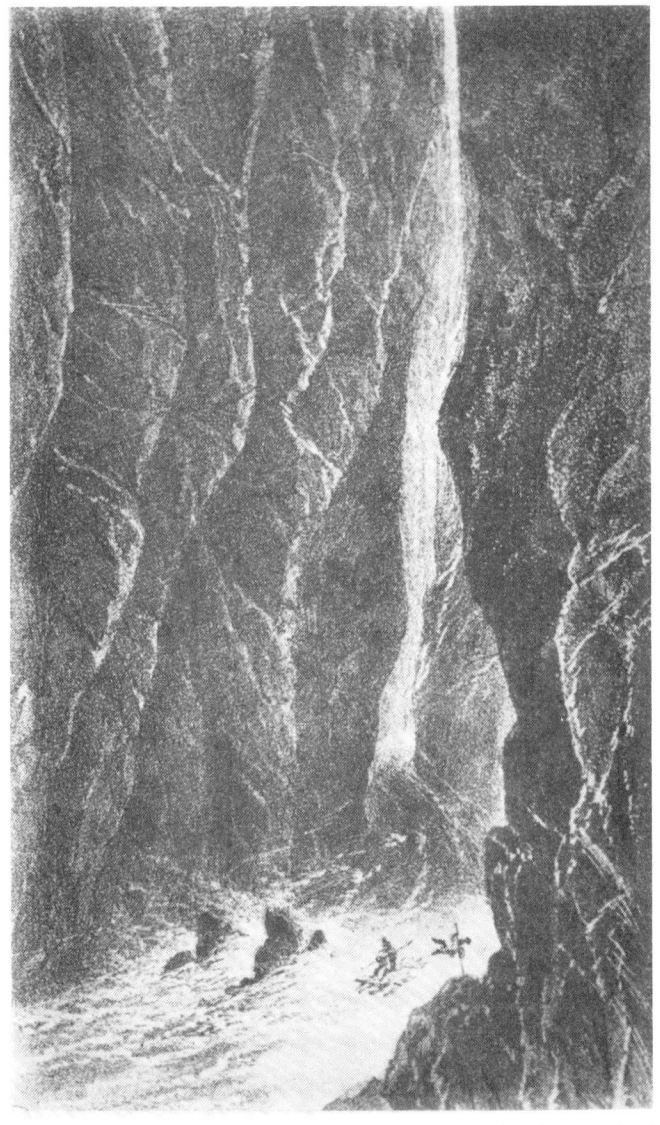

(*Reproduced from Dr. Bell's* New Tracks in North America, *London, 1870*)

IN MARBLE CANYON

James White loses his companion, George Strole, by drowning

over

IN BOULDER CANYON

(Photograph by Julius F. Stone, 1909)

Somewhere in this canyon the hapless George Strole was swept off the raft and drowned. Compare the photo with the fantastic drawing depicting the tragedy

THE HERO OF THE LEGEND 49

asked for bread and one old squaw gave me some mesquite bread. I asked for more, but she said, "Me much papooses!" and would not give me any more. This was about four o'clock in the morning on the thirteenth day when I went out with the Indians. They stole a hand-axe and a revolver from me, and one Indian wanted my coat. He was a Hualapai. He got a gun and was going to kill me, but the chief of the Utes said no.

I went down to my raft about day light and set sail again. I traveled all that day. There were no rapids to amount to anything. That night I met some Indians and traded them a revolver for a dog. They killed the dog and dressed it. I took the hind quarters. They took the forequarters. I built a fire with some coals they brought from their camp and cooked the hind quarters and went down to where my raft was and ate one quarter of the dog for my supper and laid down and slept all night, except when the fleas bit me so I could not sleep.

The fourteenth day I started out early with the raft and the quarter of the dog. I began to eat it, but I dropped it in the river. It sank like a rock. I tried my best to get it, but couldn't, so I did not have any breakfast. I sailed all that day till about three o'clock, when I struck Callville.

There were some white men there. One waded in and pulled me out of the river and helped me up the hill to the house. Jim Ferry asked me who I was.

One man gave me a shirt, another gave me a hat, another gave me a pair of pants and another a pair of shoes, until I was well fixed. I stayed there until four or five o'clock, when they had supper. They only gave me a piece of mutton, a dumpling, and one ear of corn, because I was so weak. For three or four days they issued my rations before they would let me help myself.

STANTON: How much food did you have altogether on that trip?

WHITE: When we started the first day on the river we had only half a sack of flour. We mixed it with water and ate it with a spoon. On the fifth or sixth day I got half of a mesquite bean. On the thirteenth day I got the mesquite bread from the Indians and the same evening ate one quarter of a dog.

I stayed with Jim Ferry for four weeks until I got strong. Then I went to work for him and worked for him carrying mail for about six months. Five or six weeks after I had landed at Callville I met General [William J.] Palmer [Dr. C. C. Parry] at Fort Mojave, and told him about the Grand Canyon. This is the account published in General Palmer's [Parry's] report. I had a copy of that report, but loaned it out a good many times and have lost it.

STANTON: To whom did you ever tell the story of your trip before?

WHITE: I told it first to General Palmer [Dr. Parry]

and I wrote that letter to my brother. But I have never told the full story of my trip to anyone since, up to this time when I've told it to you, excepting some conversations I had with Jim Ferry when I was at Callville. Jim wrote some account of it to the San Francisco papers and it was published there and copied into other papers.

[White refers to the letter by E. B. Grandin, published in the *Daily Alta California,* of San Francisco, Sept. 24, 1867.]

STANTON: Then you wrote that letter published in *Outing* which I wrote you about last spring?

WHITE: Yes, that's my letter. I sent and got a copy of it. I wrote that way then.

STANTON: How many big heavy rapids were there on the Colorado River that you passed over in your journey?

WHITE: In all the journey there was only one big rapid, the one with the twenty-foot fall. All the other rapids were small ones.

STANTON: Where was that one big rapid?

WHITE: It was at the end or a little above the end of the Grand Canyon.

STANTON: How many rivers or streams did you note or see running into the Colorado in the whole of your journey?

WHITE: There was only one stream coming into the river, and the water of this stream was very red. I noticed it particularly because the water was so red.

This was the Colorado Chiquito, or Little Colorado. When I passed it I did not know what it was, but afterwards General Palmer [Dr. Parry] told me it was the Little Colorado. I passed it either on the fifth or sixth day. I did not notice on the whole journey any other stream of water emptying into the river but the Little Colorado, except one stream of water running in through a hole in the rock wall one hundred feet above the river. Where this poured in the wall was all covered with moss and ferns.

STANTON: On which side of the Colorado River does the Little Colorado empty in?

WHITE: On my right side, on the right side as I was going down.

STANTON: How high were the walls of the Little Colorado?

WHITE: Not very high.

STANTON: What was the formation of the walls of the canyon? What kind of rock was it?

WHITE: The formation of the walls of the canyon was of sandstone, a light yellowish sandstone, all the way from the Green River to within 150 miles of Callville. It is all the same kind of rock, a white or yellowish sandstone.

STANTON: As a miner and a prospector you know sandstone rock, and can give an expert opinion as to its being yellowish sandstone?

WHITE: No, I'm not an expert on rock nor an expert miner, but I know yellow sandstone when I see it, just as well as I know the yellow sandstone on this ranch just outside of town. And all the rock was sandstone, white and yellowish in color, from the Green River to somewhere between one hundred and fifty and one hundred miles above Callville. There the rock is darker, a kind of greyish rock with white streaks running through it. These streaks were all on the right side of the river. I supposed these rocks were a dark greyish granite, though I did not get out to examine them.

STANTON: Where do you understand the Grand Canyon to begin?

WHITE: I understand that the Grand Canyon begins just below the mouth of the Green River.

STANTON: Where does it end?

WHITE: About one hundred and fifty or one hundred miles above Callville. I traveled four days below the canyon before I came to Callville.

STANTON: How high were the walls of the canyon through which you went?

WHITE: The walls were three or four hundred feet high. There were some higher walls, maybe, farther back. They were higher where I couldn't see them, but what I saw were three hundred or four hundred feet high, not over five hundred feet.

[Just here Mr. White's daughter interrupted to suggest that he tell me what was to her one of the most interesting incidents of her father's journey, and White complied with her request.]

I was going down the canyon on a straight part of the river when I saw a wall right across the river in front of me and no place for the river to run. I remembered the stories told of the river going through tunnels and disappearing entirely in the earth. I thought I had reached the place where I was to go through one of the tunnels or be sucked down into a great hole. When I got there the river turned sharply to the side, but there was no tunnel.

STANTON: This, then, Mr. White, is a true account of your trip down the Colorado, with, of course, many little details left out, as to where you camped, how you slept, and so forth, is it?

WHITE: Yes, this is a true account of my journey down the Colorado in 1868 [1867] and my own history, leaving out how many times the fleas bit me.

This was the end of White's story as he told it, but my investigations were not over. So I remarked, "This is certainly a very interesting story, Mr. White," and shook my head several times. White looked surprised. The following conversation then ensued.

WHITE: Well, have I told you anything wrong about the canyon?

STANTON: You have told me nothing wrong about your personal experiences. All the story of your personal experiences is undoubtedly true, and I believe every word of it. But your geography is all wrong. And what you have told about the river, the canyon, and where you traveled is entirely wrong.

WHITE: What do you know about it?

STANTON: Well, I made a railway survey down the whole river from its head to its mouth, and I am pretty well acquainted with almost every foot of it.

WHITE: Where did you make a railway survey—on top?

STANTON: No, along the bottom of the canyon next to the river.

WHITE: How could you do it? You could not walk along the bank.

STANTON: In some places you can walk along the banks. Much of the distance you cannot, but I went down the whole river in boats twenty-two feet long.

WHITE: How did you get them around the rapids?

STANTON: We ran the rapids in the boats except in some places, like the Lava Falls, where the rocks are so thick and the waves so strong that no boat, or even your raft, could last a minute.

WHITE: If you know all about it, why did you come here to ask me?

STANTON: For this reason: the story of your trip, first published in General Palmer's Report in 1868, has been rehashed every few years and republished in various magazines. The last time it appeared with your letter to your brother as published in *Outing* last April, the one I wrote you about. No one who knows anything about the Colorado River today believes your story. I believe that every word of your personal experiences is true, but I've come here to try to find out where you went, and where you actually were on the river.

WHITE: I came to the Grand River just above the mouth of the Green and sailed down all the way through the Canyon to Callville, that's where I was, and that's where I went!

STANTON: Well, let us see about it a little. You say you came to the Grand River some distance, about two miles you say, above the mouth of the Green?

WHITE: Yes.

STANTON: How did you know it was the Grand?

WHITE: Jim Baker told me. Baker knew all about that country.

STANTON: All right. Then about two hours after you set sail you passed the mouth of the Green River?

WHITE: Yes.

STANTON: How did you know it was the Green River?

WHITE: Couldn't I look right up it from the mouth?

STANTON: Yes, I suppose so. But how did you know it was the Green?

WHITE: I did not know it then, but General Palmer [Dr. Parry] told me afterwards it was the Green.

STANTON: On that portion of the Grand you traveled how high were the walls of the Canyon?

THE WRECK AT DISASTER FALLS

This highly imaginative sketch, representing the loss of the "No Name" in 1869, is reproduced from Major Powell's Report of 1875, and was allowed to be drawn four years after actual photographs of the spot had been obtained by the Major's second expedition of 1871–72

over

(*Photograph by Julius F. Stone, 1909*)

DISASTER FALLS AS THE CAMERA REVEALS IT

Compare this with the artist's conception of the "Wreck at Disaster Falls"

WHITE: About one hundred feet high.

STANTON: How high were the walls at the mouth of the Green River?

WHITE: They were not so high.

STANTON: Are you not aware, Mr. White, of the fact as shown on the Government maps that the walls of the canyon on the lower part of the Grand and at the mouth of the Green are one thousand four hundred feet high?

WHITE: No! Where we stopped at the mouth of the Green I could see out all over the mesa.

STANTON: How high were you standing above the water at the mouth of the Green when you could see all over the mesa?

WHITE: About fifty feet above the water. Just below the mouth of the Green where we entered the canyon the rock rises and stands up perpendicular, and the walls are close together, and are five hundred feet high. But at the mouth of the Green we could see out all over the mesa.

STANTON: Very well, let's go on. You say that from the time you set sail on your raft you did not pass over any rapids until you had traveled half of one night and four days below the mouth of the Green River, and that the first rapid you saw was at the end of the fourth day, when you were thrown into the river and George saved your life by pulling you out by the hair?

WHITE: Yes, there were no rapids at all for four days'

travel, and that one was the first we struck. Why, the water was so smooth that George and I sat on the raft with our feet in the water.

STANTON: Now, Mr. White, don't you know that since the time you refer to the U. S. Geological Survey has surveyed all that country and made maps of the river, and that these maps have been published? In 1889 and 1890 I made a railway survey down through all those canyons, and the reports of this work have been published long ago. In fact, commencing just about six miles below the mouth of the Green there is a stretch of the river some eighteen miles long which has a fall of 305 feet, and in which there are 57 rapids, many of them the most powerful and dangerous on the whole Colorado, with falls in some of them twenty feet and over, full of sharp rocks, and waves dashing twenty feet high, where your raft would not have lasted five seconds?

WHITE: Any man who says there are any rapids for four days travel down the Colorado River below the mouth of the Green don't know what he is talking about, and has never been on the river!

STANTON: But, Mr. White, I made an actual instrumental survey of that canyon, with transit and level and know exactly the fall of the river and the number of the rapids.

WHITE: I'll spend all the money I've got, and go over there and prove that there are no rapids there, and that I went down that river on smooth water.

STANTON: I would like to go with you if I had time. We could go over to the junction of the Grand and Green Rivers and then walk down the bank about four miles to the first big rapid. And suppose you saw that, and found, which is the fact, that these big rapids continued for eighteen miles farther down in Cataract Canyon, how would you prove that there were no rapids or that you had ever been there before?

WHITE: You could not walk along the river. Where I was in the canyon the white standstone walls were perpendicular from the water, and you can't walk along the river.

STANTON: That's just it. "Where you were," but that was *not* in Cataract Canyon just below the mouth of Green River.

WHITE: Where was I then?

STANTON: I'll come to that after a while.

WHITE: Yes, I was there, and I can prove it to you that there are no rapids there. I know a man who owns and regularly runs a steamboat on the Colorado River from the mouth of the Green three or four hundred miles down to where the rapids begin.

STANTON: That surely is interesting. I think I know the very man who told you about his steamboat. His name is Edwards—William H. Edwards.

WHITE: (Reflecting.) Yes, that's his name.

STANTON: Edwards was one of my boatmen when I

went down the Colorado in 1889 and '90. He it was who told me you were still alive and how to find you. I'll tell you about his steamboat. He ran a small boat two trips from Green River Station on the D. & R. G. R. R. down Green River one hundred and twenty miles to its mouth during 1893. In both of these he went below the mouth of the Green down the Colorado to the first rapid, where he came near losing his boat and his life, too. And the distance he ran on the Colorado was just four miles below the Green to the first rapid.

If you had gone down the Colorado River, as you say you did, after passing Cataract and Narrow Canyons, you would have had one hundred and fifty miles [Though that was the distance given in Government reports, it is actually 185 miles.] of comparatively smooth water in Glen Canyon, with a few sharp rapids. Over this stretch you might have traveled on your raft with some safety as far as Lee's Ferry.

WHITE: I don't know where that is.

STANTON: Below Lee's Ferry you would have entered another canyon where there are many terrible rapids, full of rocks and high waves. Here the president of our railroad company and two of our boatmen were drowned. Some of the rapids are so powerful that your raft, tied together with lariats, would have been smashed to pieces in no time. Then comes Grand Canyon, in the upper end of which

THE HERO OF THE LEGEND 61

there is a ten-mile stretch where the river falls two hundred and ten feet, and there are some twenty great rapids, some of them larger and more dangerous than all those above. And yet you say you passed through all these canyons on a little raft and never even saw any of these great rapids!

WHITE: I went in above the mouth of the Green and went down all the way to Callville. And there is only one big rapid, as I told you, just above the end of the canyon, one hundred and fifty or one hundred miles above Callville.

STANTON: Yes, I know those rapids, but let's look at something else.

WHITE: The rapids I went over, except the big one, the water draws in smooth from both sides to the center and just pours through, and I went over there. On the twelfth or thirteenth day I saw the flag that was put up on a rock where the first steamboat came up the river—not the flag, that was gone, but the flag pole.

STANTON: You mean Lieutenant Ives' steamboat, the "Explorer." It wasn't the first steamboat to go up there, but that's no matter. Now you have said that although you are not an expert on rock or an expert miner, you know that all the rocks of the walls of the canyon are white or yellowish sandstone, all the way to near the end of the canyon, where the rock is a dark greyish rock you supposed is granite.

WHITE: Yes.

STANTON: Now, have you never known that since 1868 the U. S. Geological Survey has published geological maps of those canyons showing just what the rock really is and that on the lower end of the Grand River the rock is a sandstone, but dark and reddish, and that for some fifty miles or more below the mouth of the Green the rock is limestone, colored generally dark and almost black by the weather? And nearly all the way through Glen Canyon for 150 miles and more, while the rock is sandstone, it is almost all red—some of it as red as fire. Then for sixty-five miles there are limestone and marble, some of the marble white, red, and variegated, and beautifully polished. Lastly, for 218 miles through the real Grand Canyon, where the walls stand over six thousand feet high, and for nearly 100 miles in two places, the walls are of black granite or gneiss a thousand or more feet high. And long stretches are limestone, but above these lower walls are piled up walls and cliffs, brown, red, yellow, green, and flaming scarlet, six thousand feet high, that can be seen from the river. And yet you tell me you went down through all these different kinds of rock and these gorgeous colors, some of the brightest and most beautiful in the world, and never saw any of them! This is certainly astonishing.

WHITE: I don't know anything about that, but where I went the rock was all white or yellowish sandstone, down nearly to the lower end of the Canyon. And the walls were three hundred or four hundred feet high. General Palmer [Dr. Parry] told me that the walls were over three thousand feet high.

STANTON: That's exactly it! Where you went, somewhere on the lower river, I suppose, but you never went in near the mouth of the Green River, and you never went through any of these great canyons at all. Just one or two more points. You have said that the Little Colorado comes into the Colorado River from the right hand side going down the river, and that the walls at the Little Colorado were not very high. Are you positively sure of that?

WHITE: Yes, I am. I saw it coming in on my right, and I know it comes in on the right.

STANTON: Have you never in all these years taken the trouble to look at a government map to find out the truth, that the Little Colorado comes in on the *left,* and not on the right, and through a canyon almost as big as the Grand Canyon, with walls nearly five thousand feet high?

WHITE: I don't know that, but I do know the Little Colorado comes in on the right, for I saw it.

STANTON: How did you know the river you saw was the Little Colorado?

WHITE: General Palmer [Dr. Parry] told me it was.

STANTON: The Little Colorado, you say, was the only stream you saw emptying into the river?

WHITE: The Green River comes in, and just below

that the San Juan comes in. We passed that. You know the Grand and the Green and the San Juan join and form the Colorado.

STANTON: How far below the mouth of the Green does the San Juan come in?

WHITE: Twenty-five miles.

STANTON: How did you know it was twenty-five miles?

WHITE: When we passed it George and I talked about it, and we both said it was about twenty-five miles.

STANTON: Did you never look it up on a map and see that the San Juan empties into the Colorado about 145 miles below the mouth of the Green and figure out the distance you traveled? You say you were floating down the Colorado at the rate of 2 or $2\frac{1}{2}$ miles an hour, and you passed the San Juan on the first day. At the average rate of $2\frac{1}{4}$ miles an hour and 15 hours a day on that smooth water you made about 35 miles a day. Hence it would have taken you just about four days to reach the San Juan. If you did reach it when you say you did, in about ten hours, you traveled the whole distance of 145 miles at the rate of $14\frac{1}{2}$ miles per hour. Pretty fast traveling on a raft on smooth water, with your feet dangling in the water, is it not? How did you get at the whole distance you supposed you had traveled on the Colorado River?

WHITE: General Palmer [Dr. Parry] told me I had

(*Photograph by R. B. Stanton, 1889*)

IN THE GRAND CANYON BELOW KANAB WASH

Here, according to Dr. Parry and James White, ". . . the route was more open"

traveled five hundred and fifty miles.

STANTON: You told your story first to 'General Palmer'?

WHITE: Yes.

STANTON: Were any other members of 'General Palmer's' party present.

WHITE: No, sir.

STANTON: Now think, and try to recollect. Did you see any other members of 'General Palmer's' party at any time at Hardyville, or anywhere and tell them this story in the winter of 1867 and 1868?

WHITE: No, sir, no one but General Palmer [Dr. Parry] himself.

STANTON: Just one more point, please. When you went from Silverton and struck the San Juan—

WHITE: We went down the Mancos to the San Juan.

STANTON: Very well, you went down the Mancos and came to the San Juan and you crossed the San Juan and went right on two days to the camp in the canyon near where Baker was killed?

WHITE: Yes, sir.

STANTON: And then twelve or fifteen miles down the canyon to the Grand River?

WHITE: Yes, sir.

STANTON: Now (drawing another sketch in note book and pointing to it) here is the Grand, here is the San Juan, and here is the Colorado. You came here, down the Mancos to the San Juan, and crossed over the San Juan, you say? Now where did you go then?

WHITE: If that is the Grand, we went up there to the Grand.

STANTON: But you crossed over the San Juan coming to it from the north, and so got on the south side. How did you get from there to the Grand River? Did you cross back over the San Juan a second time?

WHITE: No, we only crossed the San Juan once.

STANTON: Are you positive?

WHITE: I am positive.

STANTON: Think carefully. You crossed the San Juan after coming down the Mancos and then in two days went to the Grand River, and you are absolutely positive that you never crossed the San Juan river but once?

WHITE: Yes, sir, I am positive we only crossed the San Juan once.

STANTON: Now, Mr. White, I will tell you what I think of all this. First, be sure to understand I believe every word you say as to your personal experiences. I do not doubt your word. As far as you know, you have told the truth. You were on a raft. You floated down part of the Colorado, but not where you thought you did, or rather where you were *told* you did. You have been misled by others. You were picked up from the raft at Callville in September 1867 in a pretty demoralized condition. That is a proved historical fact.

THE HERO OF THE LEGEND

When you were on the raft I think I know where you saw the white walls and then the dark walls. You saw the one river coming in on your right—I think I know what river it was—and you finally reached Callville. But you never went to the Grand River, and you never floated on a raft through Cataract, Glen, Marble, or Grand Canyon. The fact is that from the time you struck the San Juan River at the mouth of the Mancos River, *you were lost.* You did not know where you were, or where you went.

WHITE: Maybe I was!

STANTON: I will tell you two more things, and these in utmost kindness. I intend to defend you against the charges made, but do you know that most of the people who know anything about the Colorado River think you are (and you are so mentioned in several books and pamphlets) the biggest liar that ever told a story about the Colorado? But I do not think so at all. You have told the truth as far as you knew, that is, as to your personal experiences. And further it has been believed by some very distinguished men that you murdered your two companions and told your story in 1867 to cover up your crime.

WHITE: I didn't kill them!

STANTON: No, I know you didn't, and now that I know the truth I am going to defend you from both charges.

WHITE: You are going away tomorrow?

STANTON: Yes, I must go on the early train.

WHITE: No, don't go. I like you. Stay another day, and I'll put up my horses in the barn and we will spend the day together and talk it all over.

But I could not stay. So we bade each other goodnight with a hearty shake of the hand.

AFFIDAVIT:

STATE OF COLORADO,
COUNTY OF LOS ANIMAS, } ss
CITY OF TRINIDAD:

I, Roy Lappin, being duly sworn, do depose and say as follows: I am a public stenographer doing business in the City of Trinidad, Colorado. On the 23rd day of September, 1907, I was employed by Robert B. Stanton of the City of New York, to take down in shorthand an interview which he had with James White, residing at No. 401 Short Avenue, in the city of Trinidad, Colorado, and on the evening of said day did go with the said Stanton to the residence of the said White and there took down in shorthand said interview, and furnished the said Stanton a typewritten copy of my notes, and further, the foregoing twenty-three pages contain a true and correct transcript of that said interview, together with the conversation had between the said Stanton and the said White, with certain corrections in grammar and other minor details, which in no way change the meaning or intent of the interview or statements made by the said White.

      (signed) Roy Lappin

Sworn to before me, William H. Humphreys, this the 9th day of November, 1907.

      (signed) William H. Humphreys

CHAPTER IV

RAFT JOURNEYS—IMAGINARY AND REAL

THE Dawson pamphlet of 1917, which attempts to transfer from Major Powell to James White the credit of being first to go through the Grand Canyon, consists, I must say, principally of unfounded assertions and unsubstantiated conclusions. Were all of Dawson's statements founded on fact, I should not have left a leg to stand on. But it is easily possible to take Dawson's most important assertions one after the other and demonstrate their utter fallacy.

"Numerous contemporaries," Dawson declares, "bear testimony to the fact that at the time he emerged from the river he asserted that he had come through the canyon"—meaning, of course, the 500-mile canyon. However, Dawson produces not one particle of testimony that White ever made such an assertion within months of his arrival at Callville. Hardy, McAllister, Ballard, Tillman, and others become convinced that White had traversed the Big Canyon, and *they* said he had. But they give no direct testimony that they ever heard White himself say so.

Dawson says again, referring to the testimony of a list of witnesses supporting the tale he is championing, "In the main this consists of records prepared by men who saw and conversed with White very soon after the conclusion of his trip."

Actually, of the twenty witnesses thus put forward, only four saw and talked with White soon after his arrival at Callville. They were W. H. Hardy, Charles McAllister, Dr. C. C. Parry, and a man named Ballard. Other principal witnesses named are General William J. Palmer (Dr. Parry's chief of survey), Dr. William A. Bell, the physician of the survey, and Major A. R. Calhoun, attached to Palmer's party as correspondent for the *Philadelphia Press*. General Palmer did not meet James White until a year and a half later, in 1869. Dr. Bell did not meet White until 1917. And Major Calhoun never saw him at any time. Yet it is upon these three, and principally Major Calhoun, that Dawson relies, aside from Dr. Parry.

Gen. Palmer not only believed the Parry story as applied to the whole 500 miles of the great Canyons, but he fought for it with almost his last breath. He went farther than most of White's advocates. He investigated it as far as he thought necessary, and he discussed it with Major Powell, Col. R. C. Clowry, President of the Western Union Telegraph Company, and certainly with many others, and finally with me in a number of letters. Probably all the information on the subject known at that time was in his possession, or easily available to him. In February 1908 I sent Gen. Palmer a very full copy of all data I had gathered, including the stenographic report of my interview with White, so that he, for one, was not lacking in information, so far as he would receive it.

With all this in view, if Gen. Palmer had only argued the case in the manner that Dawson and others have done (in fact he did this, and he did it better than any one else has) and then given his opinion and judgment, it would have been difficult to refute, coming from such a man. He went farther, however, and based his final judgment upon the assumption of the accuracy of the description of the whole 500 miles of canyons as it was supposed to have been given by White.

Gen. Palmer argued, mind you, in 1906, 1907, and 1908 that White made the whole 550-mile raft journey down the Colorado in 1867 because before that time no one had known anything of the geology of the interior of the canyons, and that since White gave an absolutely correct description of the geology of the whole distance, he could not be mistaken as to where and how far he had traveled!

In almost any large library one may find the complete geological works of Newberry, Powell, Dutton, Walcott, and others on the Grand and other canyons of the Colorado. These authorities affirm unequivocally the existence of dark limestone in Cataract Canyon, one hundred and eighty-five miles of mostly red sandstone in Glen Canyon, some sixty-five miles of red and variegated colored limestone and marble in Marble Canyon, and mile upon mile of black and red granite in the Grand Canyon. And nowhere in these geological reports is any "long continued" "white sandstone" shown, for it does not exist.

When in 1908 I called his attention to the fact that no part of White's description was applicable to the canyons above the Grand Wash Cliffs, he still stuck to the alleged geological proof coming only from White, ignoring and thereby denying the truth of the many accepted scientific reports to which I had called his special attention. This astonishing condition of affairs can be accounted for in only one way: Gen. Palmer was so absolutely prejudiced in favor of Parry's reliability that he could not and would not see any truth in anything or anybody that contradicted the White story. It would seem, in fact, that the good General was as negligent of duty as Dr. Parry, and as stubborn as James White himself.

From the Parry story have sprung a great many other accounts of the marvelous raft voyage. As thrillers some of them have been quite notable. Their claim to historical accuracy is, however, negligible in the extreme. By all odds the most sensational and romantic accounts of James White's adventure came from Major A. R. Calhoun, a newspaper correspondent who wrote his tales with the aid of Dr. Parry's brief notes and his own boundless imagination.

The first of Major Calhoun's efforts to do justice to the opportunities afforded by James White appeared in Dr. Bell's volume, *New Tracks in North America,* published in London and New York in 1870. In another account, appearing in a volume entitled *Wonderful Adventures,* published in Philadelphia about 1875, Major Calhoun gives his fancy even freer rein, and illustrates his wild tale with drawings vastly more absurd than those reproduced in Dr. Bell's book.

The story as carried in Dr. Bell's book is concluded with a verbatim presentation of Dr. Parry's nine conclusions or additions to the previously existing geographical knowledge of the region. Though he says that he met White personally, it is certain that Calhoun did not do so, but depended wholly upon Dr. Parry's notes, rumors, and a facile imagination. Perhaps the only satisfactory way to convey an impression of the florid, dime-novel style of Major Calhoun's wonderful yarn is to quote some of it as we find it in *Wonderful Adventures.* Let us begin with the Indian attack:

Early next morning they breakfasted, and began the ascent of the side cañon, up the bank opposite to that by which they had entered it. Baker was in advance, with his rifle slung at his back, gaily springing up the rocks, towards the table-land above. Behind him came White, and Strole with the mules brought up the rear. Nothing disturbed the stillness of the beautiful summer morning but the tramping of the mules, and the short, heavy breathing of the climbers. They had ascended about half the distance to the top, when stopping for a moment to rest, suddenly the war-whoop of a band of savages rang out, sounding as if every rock had a demon's voice. Simultaneously with the first whoop a shower of arrows and bullets was poured into the little party. With the first fire Baker fell against a rock, but, rallying for a moment, he unslung his rifle and fired at the Indians, who now began to show themselves in large numbers, and then, with the blood flowing from his mouth, he fell to the ground. White, firing at the Indians as he advanced, and followed by Strole, hurried to the aid of his wounded leader. Baker, with an effort, turned to his comrades, in a voice still strong, said, "Back, boys, back! save yourselves, I am dying." To the credit of White and Strole be it said, they faced the savages and fought, till the last tremor of the powerful frame told that the gallant Baker was dead. Then slowly they began to retreat, followed by the exultant Indians, who stopping to strip and mutilate the dead body in their path, gave the white men a chance to secure their animals, and retrace their steps into a side cañon, beyond the immediate reach of the Indians' arrows. Here they held a hurried consultation as to the best course they could pursue. To the east for 300 miles stretched an uninhabited country, over which, if they attempted escape in that direction, the Indians, like bloodhounds, would follow their track. North, south, and west was the Colorado, with its tributaries, all flowing at the bottom of deep chasms, across which it would be impossible for men or animals to travel. Their deliberations were necessarily short, and resulted in their deciding to abandon their animals, first securing their arms and a small stock of provisions, and the ropes of the mules. Through the side cañon they travelled, due west, for four hours, and emerged at last on a low strip of bottom land on Grand River, above which, for 2,000 feet on either bank, the cold, grey walls rose to block their path, leaving to them but one avenue for escape—the foaming current of the river, flowing along the dark channel through unknown dangers.

Then comes the drowning of Strole and lone terror for White, who even envies his companion in now being beyond fear:

But about three o'clock on the afternoon of the 28th they heard the deep roar as of a waterfall in front. They felt the raft agitated, then whirled along with frightful rapidity towards a wall that seemed to bar all further progress. As they approached the cliff, the river made a sharp bend, around which the raft swept, disclosing to them, in a long vista, the water lashed into foam, as it poured through a narrow precipitous gorge, caused by huge masses of rock detached from the main wall. There was no time to think. The logs strained as if they would break their fastenings. The waves dashed around the men, and the raft was buried in the seething waters. White clung to the logs with the grip of death. His comrade stood up for an instant with the pole in his hands, as if to guide the raft from the rocks against which it was plunging; but he had scarcely straightened himself, before the raft seemed to leap down a chasm, and amid the horrible sounds White heard a shriek that thrilled him. Turning his head, he saw through the mist and spray the form of his comrade tossed for an instant on the water,

RAFT JOURNEYS—IMAGINARY AND REAL 77

then sinking out of sight in the whirlpool.

White still clung to the logs, and it was only when the raft seemed to be floating smoothly, and the sound of the rapids was behind, that he dared to look up; then it was to find himself alone, the provisions lost, and the shadows of the black cañon warning him of the approaching night. A feeling of despair seized him, and clasping his hands he prayed for the death he was fleeing from.

And this is the way the Major warms to his task when he comes to relate White's experience in being caught in the big eddy at what he was led to believe was the mouth of the Little Colorado:

When we reached the mouth of the latter stream the raft suddenly stopped, and swinging round for an instant as if balanced on a point, it yielded to the current of the Chiquito, and was swept into the whirlpool. White felt now that all further exertion was useless, and dropping his paddle, he clasped his hands and fell upon the raft. He heard the gurgling waters around him, and every moment he felt that he must be plunged into the boiling vortex. He waited, he thinks, for some minutes, when, feeling a strange swinging sensation, he looked up to find that he was circling round the whirlpool, sometimes close to the vortex and again thrown back by some invisible cause to the outer edge, only to whirl again towards the centre. Thus borne by the circling waters, he looked up, up, up through the mighty chasm that seemed bending over him as if about to fall in. He saw in the blue belt of sky that hung above him like an ethereal river, the red-tinged clouds floating, and he knew the sun was setting in the upper world. Still around the whirlpool the raft swung like a circular pendulum, measuring the long moments before expected death. He felt a dizzy sensation, and thinks he must have fainted; he knows he was unconscious for a time, for when again he looked up the walls, whose rugged summits towered 3,000 feet above him, the red clouds had changed to black, and the heavy shadows of night had crept down the canyon. Then, for the first time, he remembered that there was a strength greater than that of man, a Power that "holds the ocean in the hollow of His hand." "I fell on my knees," he said, "and as the raft swept round in the current, I asked God to aid me. I spoke as if from my very soul, and said, 'O God! if there is a way out of this fearful place, guide me to it!' " Here White's voice became husky, as he narrated the circumstance, and his somewhat heavy features quivered, as he related that he presently felt a different movement in the raft, and turning to look at the whirlpool, saw it was some distance behind, and that he was floating down the smoothest current he had yet seen in the canyon.

No wonder Dr. Wm. A. Bell admitted in later years that Calhoun embellished the story considerably! As for Dr. Bell himself, it is apparent that he at first accepted the White story in good faith because it came to him on the authority of his friends Parry and Calhoun, making no particular investigation into the subject matter. It was a good story, and Dr. Bell considered it a desirable addition to his book.

Unlike General Palmer, who stubbornly refused to

consider the case against White, Dr. Bell changed his mind on more than one occasion in the years to come as new material bearing on the whole affair was brought to his attention. After I had had my interview with White in 1907 I wrote to Dr. Bell and gave him a synopsis of my findings up to that time. In his reply he remarked of the Calhoun version: "Major Calhoun was for some years a newspaper correspondent of the *Philadelphia Press,* and it is not improbable that he seized upon the information furnished by Dr. Parry as a good foundation for a thrilling romance. Even, today I cannot read his account without being greatly impressed with it. I am not, however, surprised to hear from you that it can only be said to some extent to be founded on facts."

In 1917 I sent Dr. Bell a copy of the Dawson pamphlet, along with my discussion of Dr. Parry's notes. The pamphlet made a marked impression on him, as is shown in one of his letters to me: "I read Mr. Dawson's pamphlet with great interest. It seems to remove every doubt, if any existed, as to whether White did pass through the canyon or not." And of Dr. Parry he observes in the same letter, "Dr. Parry I consider a very responsible and trustworthy observer. He was left behind by Gen. Palmer for the express purpose of interviewing White. His report of the interview can be relied upon absolutely . . . Dr. Parry's [account] is beyond suspicion."

Several weeks later Dr. Bell, returning from a trip

to California, stopped off at Trinidad, Colorado, and spent a day with James White, then eighty years old. It was shortly after this that I submitted to him my criticism of the Dawson pamphlet. Several months later he wrote: "You have taken infinite pains to show how untrustworthy is the evidence that White passed through the whole length of the canyons of the Colorado. . . . I wish you had told us more in detail of the river for that sixty miles above Callville. . . . I shall tell my friends in future, if they are interested in the question, that I do not think White did pass through the series of canyons he was supposed to have traversed, but that sufficient evidence is still lacking, in my opinion, to definitely fix his starting point."

Independent proof of the exact starting point it is obviously impossible to produce. If it were possible, there would be no dispute whatever. The matter of White's starting point, which Dawson mentions as "a new question," has been argued ever since the raft voyage. The discussion began at Callville, and in the next fifty years at least seven separate and distinct points were named as the place where White and Strole launched their raft. They are as follows:

      1867—Mouth of the San Juan—White and friends at Callville.

      1868—Thirty miles up the Grand—Dr. Parry's assertion.

      1906—Lower end of Cataract Canyon—Gen. Palmer's suggestion.

(*Photograph by Hillers, courtesy U. S. G. S.*)

LAVA "FALLS"

Another reason why it is hard to believe the Colorado-River-on-a-raft yarn

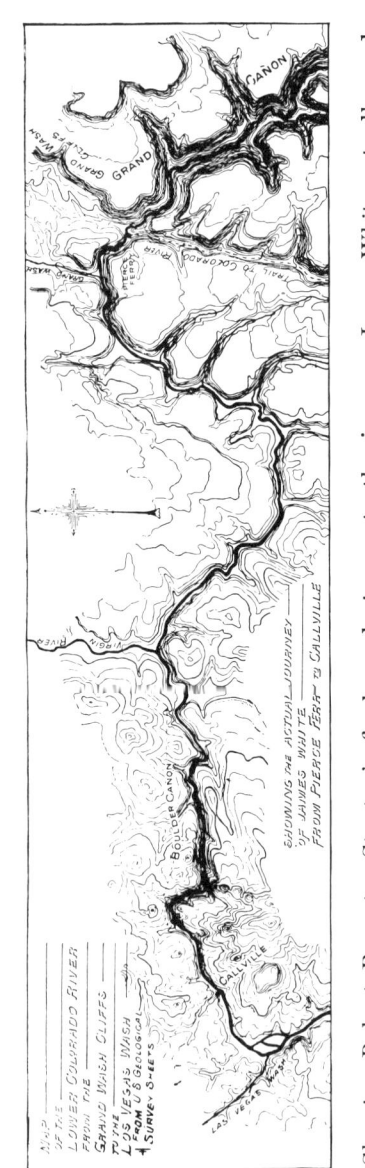

Showing Robert Brewster Stanton's final conclusion as to the journey James White actually made on his fabled raft, or rafts

1907—Two miles above the "head of the Canyon"—White.
1916—Five miles below the "head of the Canyon"—White.
1917—Nine miles below the "head of the Canyon"—White.
1917—Forty miles above mouth of the San Juan—Dawson.

All we know positively is that White was on the Colorado at Callville. But since Dr. Bell now admits that White did *not* go through the series of big canyons, then he must believe, as an unavoidable corollary, that he began his river journey somewhere between the lower end of the Grand Canyon and Callville, whether at Pierce's Ferry, as I think, or at Scanlon's Ferry, or the gulch lying between them being of little importance. In effect, therefore, Dr. Bell has changed his opinion of the White story completely.

Dawson argues that White furnished valuable information because of which future scientific explorations might be made easier. "Is not this fact that White supplied information on which to base such calculations," he demands, "an item in his favor?" Referring bluntly to the leader of the 1869 expedition, he says: "Major Powell is entitled to commendation for not entering upon his venture without knowing that the river was free from unsurmountable obstacles. That fact had been demonstrated by White." And further: "Knowing, then, that the Canyon had been navigated from end to end; knowing that it no longer was entirely untraveled, and knowing especially that it was without high water falls, Powell might well enter on his survey, appreciating the danger of the undertaking, but still willing to take the risk for the further advancement of the world's wellfare and the promotion of his own fame."

The implication here is inescapable: Major Powell did not determine upon his river venture until after James White had led the way and given assurance that in the whole course there was but one bad rapid. That is making the Major play second fiddle with a vengeance!

Here again the facts are decidedly not in support of Dawson's assertions. Early in the summer of 1867 Major Powell, with his group of college students, went into Middle Park, Colorado. That summer, months before James White had ever seen the Colorado at any point, and nearly a year before Powell could possibly have heard of White's adventure, the Major entered into an agreement with Jack Sumner to explore the canyons of the Colorado the following spring.

When, some time in 1868, Powell read in the newspapers or was told by Samuel Bowels of White's exploit, there was at that time no reason for doubting its truth. Knowing how bold frontiersmen were, he undoubtedly accepted the story. What emotions the Major may have experienced on hearing that someone else had accomplished the feat which he and Sumner had planned and determined upon the year before, and especially on learning that in all its length the river contained only one big rapid, it is impossible to say. To some men the knowledge that the river was safe would have given great satisfaction, but for Major Powell to learn that the perilous journey he had determined upon was to be nothing but a quiet float down a smooth stream must have been quite a disappointment. I can imagine that during his rest in his peaceful camp at the junction of the Grand and the Green the good Major's belief in White's description of what was to follow throughout the whole length of the Colorado had reached its high water mark. But when he had gone about four miles into Cataract Canyon, he began to tumble down fifty-seven big rapids (the cataracts which give their name to the canyon). I can perfectly well understand why the Major then and there denounced White as "the biggest liar that ever told a tale about the Colorado River," and how after passing through the Marble and Grand Canyons, with his experiences in "Sockdologer," "Separation," and the hundreds of other big rapids, and finally arriving in safety at the Virgin River, he still called White a "monumental prevaricator"—thanks to Dr. Parry!

Nevertheless, in one place Dawson makes the statement that "in at least one instance, the *achievement* (not the account) received *official recognition* from an organized body of scientific men." This refers to the publications of Dr. Parry's paper by the Academy of Science of St. Louis. No such official action was ever taken. The mere printing of the paper in the Transactions carried no endorsement with it. Neither the Academy as a body, nor any member of the Academy at the request of the Academy ever investigated the truth of the subject under discussion.

I shall not attempt to reply to all of Dawson's statements regarding the rapids he claims White passed over. Many of them cannot be found in the records he or any one has published. They come only from Dawson's imagination, to suit his arguments. On this one point White has been clear, distinct, and truthful for fifty years, never once varying in his story in even the smallest degree from his first letter to his brother in 1867 to his last account in 1917. He has told at various times of many rapids, whirlpools, and eddies over which he passed in that section of the river and the canyons where he went, but he has never written or told in all this time of more than the one big rapid over which his raft tumbled "on the eleventh day."

Looking over the original notes of my survey where I recorded the number, position, and character of each and all the rapids (from the junction of the Grand and Green five hundred and twenty to the end of the Grand Canyon as series "A," and below the Grand Canyon as series "B"), records made for railway pur-

RAFT JOURNEYS—IMAGINARY AND REAL 85

poses, not to check up White, I find that from Grand Wash Cliffs to Callville I noted thirty-one rapids, all of the character White described to me. Many of these were most insignificant, compared with those above. Others are noted as "light," "brisk," "sharp," and *one* and *only one* in that whole distance, as *"heavy rapid,* six to eight feet fall." This one rapid is near the end of one of the small canyons White went through, but not at the extreme lower end of the Boulder. But this again merely demonstrates White's utter inability to orient himself. Undoubtedly this was White's one big rapid.

Let us see how White's description of the river and its rapids in all the journey agrees with the conditions that actually exist in the great canyons *above* Grand Wash Cliffs. In that upper section of about 500 miles there are some 520 rapids, cataracts and falls, by my count, both large and small. Somewhere near 300 of them are large, and many of them much larger, more powerful, and with greater, and steeper falls, than the one big rapid which White described. At one place in Cataract Canyon (which, by instrumental measurement, has a fall of 304 feet in eighteen and a half miles) there are 57 big, powerful rapids bunched together. At another, in the Grand Canyon (with a fall of 165 feet in ten miles), there are some of the biggest, steepest, and rockiest rapids on the river, besides the rest of the 300 big ones scattered through Cataract, Narrow, Marble, and the Grand Canyons.

Some of these have individual drops of 30 to 35 feet.

In the face of these facts, proved by instrumental measurements, official surveys, and the reports and testimony of a score of men who have navigated all of them, some of whom have had their strong wooden or steel boats smashed to pieces against the rocks, we are asked to believe that White floated on a little raft tied together with ropes over the whole 300 and never saw or felt any of them except one!

Dawson says that White, entering at the head of "the canyon," "found himself locked within its walls," and he continued to the end because he discovered "no means of escape from its compelling embrace." What is the truth about it? If White had entered the canyon near the junction of the Grand and the Green, he would have found difficulty in getting out while passing through Cataract Canyon. There, however, he was not looking anxiously for a way of escape. He and Strole were having a regular summer picnic, sitting on the raft with their feet dangling in the quiet, smooth water.

Reaching a short distance below the Dirty Devil River, if they had wished to escape from that compelling embrace they could have found fifty easy ways in the next 185 miles, anywhere along the broad flats of Glen Canyon. In many places broad, easy side-canyons lead out—Crescent Creek, White Canyon, Dandy Crossing, Trachyte, Red Canyon, Tickaboo

Creek, the Crossing of the Fathers, and many others, down to Paria Creek (Lee's Ferry), that broad, open break-down of the walls. At the head of the Grand Canyon proper occurs still another. And in both Marble and Grand canyons also there are a number of side gorges that would have been worth trying.

Dawson argues against the criticisms of White's inaccuracies in describing the interior of the great canyons where he is alleged to have been on his raft as mere faults of detail and so of no consequence. Many of his faults of detail and minor contradictions even, are of course of no importance, and not worth wasting breath upon. Others, when analyzed, become significant in the effort to decide the main question as to whether he did or did not make the alleged raft voyage on the Colorado.

If White had been able to describe correctly only a few prominent features of the interior of the great canyons, though he might have varied in some minor details, this would have been proof positive that he had been at those points, and in case they extended well up in the Grand and Marble Canyons this again would have been convincing proof that he had passed through the big canyons, for the reason that no one had done so before his time from whom he could have gotten the information.

One example will serve to illustrate how plainly erroneous were some of White's positive assertions. At the time I interviewed him White told me, "On the twelfth or thirteenth day I saw . . . the flagpole," the one that was put up by Lieut. Ives on Fortification Rock. And in his 1917 statement he says: "While floating in the moonlight I saw a pole sticking between two large rocks, which I afterwards learned the Government had placed there some years before, as the end of its journey." That pole and what he was told about it are just the things such a man would naturally observe and distinctly remember. But when he and others use such facts in drawing conclusions, and put them forward as proof that he made the whole 550-mile journey on his raft, their arguments fall flat.

There is no reason whatever to doubt that White saw that flagpole as he says, saw it both in daylight and by moonlight, but he did *not* see it on the twelfth or thirteenth day of his raft trip. According to Parry's and Dawson's estimates of the distances White traveled daily, he was on those days somewhere between 55 and 105 miles up the river from where that pole was situated. However, White most certainly did see that flagpole by daylight and probably by moonlight also, and while floating on the Colorado. The solution is this:

The location of the flagpole was five miles below Callville. After recovering his health at Callville, White floated down stream from there by boat to Hardyville and Fort Mojave, probably several times,

RAFT JOURNEYS—IMAGINARY AND REAL 89

and one or more times it may have been by moonlight. Then it was that he saw the pole and was told what it was. And ever afterwards he has clearly remembered the fact, but being almost devoid of any reasoning or analytical powers, he fails to separate the boat journeys *below* Callville, from the raft journey *above*. That he did not see that pole while on his raft, or even any other flagpole above Callville, by moonlight on the nights he is said to have been on his raft is proved by astronomy. Nor can we accept White's statement that on one occasion when he was asleep on an island the moonlight was so bright that he was awakened at three o'clock in the morning, thinking it was daylight. For on the nights of August 25 to September 8, 1867, the dates given by Dr. Parry, the moon was not visible after about midnight at any time, and that was on September 8th, when the moon shone as late as 12:41 A. M. The almanac settles this matter with no room for argument:

TIME OF RISING AND SETTING OF MOON IN 1867

	AUG.	RISE	SET
AT JUNCTION OF GREEN AND GRAND RIVERS	25″	1.05 A. M.	3.35 P. M.
	26″	2.06 A. M.	4.31 P. M.
	27″	3.10 A. M.	5.23 P. M.
About Lat. 38 N. Long. 110 W. of Greenwich.	28″	4.18 A. M.	6.08 P. M.
	29″	5.27 A. M.	6.49 P. M.
	30″	6.35 P. M.	7.26 P. M.
	31″	7.40 P. M.	8.02 P. M.

	AT CALLVILLE, NEVADA, ABOUT Lat. 36 N. Long. 114 30 W. of Greenwich.	SEPT.	RISE	SET	
		1	8.38 A. M.	8.42 P. M.	
		2	9.41 A. M.	9.16 P. M.	
		3	10.41 A. M.	9.53 P. M.	
		4	11.40 A. M.	10.30 P. M.	
		5	12.34 P. M.	11.11 P. M.	
		6	1.27 P. M.	11.54 P. M.	Sept.
		7	2.18 P. M.	12.41 A. M.	8th
		8	3.02 P. M.	1.30 A. M.	9th

After summarizing the attacks which have been made on the White story, Dawson remarks, "Necessarily this criticism is based largely on the ground of improbability." As a matter of fact, this was largely so up to the time when I began my extended investigations into the subject, commencing with my interview with White in 1907. To those of us—and our name is not legion—who have actually traveled the whole length of the tempestuous Colorado and passed at the bottom of the gorge through all of its awe-inspiring canyons the James White story had always seemed a myth. And yet it was most difficult entirely to disprove it, for it was founded on a grain of truth, that is, that White was actually rescued from a raft while floating down the lower Colorado below all of its great canyons.

The improbability of such a raft journey as has been attributed to White was clearly expressed in 1906 by Julius F. Stone, long conversant with the Colorado River and leader of his own successful river

RAFT JOURNEYS—IMAGINARY AND REAL

expedition in 1909. In a letter to Mr. R. C. Clowry, of New York, Mr. Stone writes:

I am very much inclined to the opinion expressed by Major Powell and Mr. Stanton that it would be impossible for one or two men to pass through the canyon on a raft of those 8 inch cottonwood logs 10 feet long and live. The specific gravity of ordinarily dry cottonwood is about $\frac{1}{2}$ and the cubical content of 4 logs 8 inches in diameter and 10 feet long is about $10\frac{1}{2}$ feet, which would give a buoyant effect of about 425 lbs. at the point of total immersion. I presume it would be safe to estimate the weight of each of the two men at 175 lbs., which would leave about 75 lbs. available for their arms and provisions up to the extreme limit of submergence; but the specific gravity of the raft would increase quite rapidly by reason of the dry wood absorbing water up to the limit of saturation, thereby reducing very quickly the small margin between the weight of the men and the point of total immersion of the raft. It will occur to you at once that the moment of stability of such a raft 10 feet long and two feet wide would be excessively small, so much so that it would tax the ability of any but the most expert "log drivers" to pass through even ordinarily rough water on such a craft without disaster.

Coming down the Grand River, as they state, they must have passed through Cataract Canyon before reaching the much stiller water found in Glen Canyon immediately below, and presuming that they successfully navigated the turbulent waters of Cataract, they must have been either very much more or less than human if they failed to take advantage of the opportunity afforded to leave the river at any of the several places where this might easily have been done throughout the entire length of the Glen Canyon, there being three points, namely Dandy Crossing, the Crossing of the Fathers, and Lee's Ferry where they could have left the river and gone in either direction.

It would be idle for me to deny *in toto* that Mr. White did go through the entire canyon as he claims, but I think the probability of his having done so is about in the ratio of one to infinity.

But the fact is that some of us have looked into the matter more or less thoroughly. It was but natural for us to be incredulous. Still, our saying it was impossible did not necessarily make it so. Now it is no longer a matter of probabilities. From my own investigation I no longer merely say I believe it could not happen—I say it *did not* happen.

Summing up the whole thing and stating my earnest convictions with regard to the alleged voyage and the real voyage of James White, the following are my conclusions: The statements by White as to the nature and height of the canyon walls which he saw, the nature of the river, the kind of rapids over which he passed, and the number of big rapids he encountered—all these are faithful accounts of genuine personal observation and experience.

And each of them, even as regards the unknown river which he saw coming in on his right (the Virgin, which Dr. Parry erroneously insisted was the Little

RAFT JOURNEYS—IMAGINARY AND REAL

Colorado), is distinctly and absolutely true when applied to that portion of the Colorado River between Pierce's Ferry and Callville. At the same time, not a single one of them is even approximately true when applied to that part of the Colorado above the Grand Wash Cliffs.

And so, from his own testimony, I conclude that James White never passed through a single mile of the canyons of the Colorado River above Grand Wash Cliffs, but that he did float on a raft or rafts on that river in the year 1867 a distance of sixty miles from a point near the Grand Wash to Callville, Nevada, where he was stopped and taken off his raft. To Major John Wesley Powell, leader of the expeditions of 1869 and 1872, therefore, belongs beyond the shadow of a doubt the honor and distinction of having been the first conqueror of the Colorado River.

PART II

THE AFFAIR AT SEPARATION RAPIDS

CHAPTER I

MAJOR POWELL AS HISTORIAN

IN discussing the expedition of 1869 it is my object to place on record certain things which have hitherto never been told adequately. Some of these facts I published as early as 1889, and some have been discussed by other writers. No new glory can be added by any eulogy from me to the gallant Commander of the first real exploration of the river and the canyon, but I offer my meed of praise in all sincerity.

To John Wesley Powell belongs all the glory and renown possible for his first trip through the unknown canyons of this wonderful river. Starting as he did with only a scientist's conjectures of what might be before him required far greater bravery than to follow him. I had a very much greater admiration for his determination and that of his companions after I had traveled over their route and seen something of the hardships and dangers they had to undergo. At that time no one was more ready to render homage to the gallant soldier and his companions than I.

In various publications and on perhaps fifty lecture platforms, I have paid my tribute to Major Powell since I completed my journey through the whole of the Grand Canyon. And now, after a lapse of over thirty years since my first work on the Colorado began, my admiration for the daring feat of the brave little band who started and the six who completed that exploration of 1869 has risen immeasurably. Whatever may have been accomplished since that time, whoever may have traversed the great canyons or braved the terrors of the five hundred and more cataracts and rapids of the river, it is but following that little band of '69 who blazed the way. True, the river is the same as it was fifty years ago—yes, perhaps five hundred years ago. The power of the current is the same, and the fall from the head to the mouth of the Grand Canyon is still as great as it was then, and the dangers of navigating a boat over and through the lashing waves of the cataracts is as great today as ever, and no skill, no experience, can enable one to foresee what a single cross-current wave may do. Still, the terror of the "Great Unknown" was forever removed when Major Powell passed through the lower gateway at the Grand Wash Cliffs. It is therefore one of my greatest pleasures and I consider it an honor to record my tribute to the man whose courage and determination forever dispelled that mystery.

After spending nearly a year in an engineering journey through the depths of the stupendous chasms of that tempestuous river, experiencing near starvation and witnessing the death of three of my companions in its pitiless rapids, I am perhaps as well fitted as most men to appreciate the bravery of the men who solved the problem of those then unknown depths. Because of actual experience in those same canyons I can all the better appreciate what thoughts and emotions must have been his when of the time he was about to enter the Grand Canyon itself the dauntless Major wrote:

"We are now ready to start on our way down the Great Unknown. Our boats, tied to a common stake, are chafing each other as they are tossed by the fretful river. We have not a month's rations remaining. The flour has been resifted, the spoiled bacon has been dried and the worst of it boiled, the few pounds of dried apples have been spread in the sun and reshrunken to their normal bulk, the sugar has all melted and gone on its way down the river; but we have a large sack of coffee.

"We are three-quarters of a mile in the depths of the earth and the great river shrinks into insignificance as it dashes its angry waves against the walls and cliffs that rise to the world above; they are but puny ripples and we but pigmies, running up and down the sands, or lost among the boulders.

"We have an unknown distance yet to run; an unknown river yet to explore. What falls there are, we know not. What rocks beset the channel, we know not. We may conjecture many things. The men talk as cheerfully as ever; jests are bandied about freely this morning, but to me the cheer is sombre, and the jests are ghastly."

Taking into consideration the conditions that surrounded him at the time, Major Powell's work and that of his companions in 1869, to my mind, stands out as one of the bravest exploits ever known anywhere. It is not my intention to rewrite the story of the exploration of 1869, for Major Powell's famous Report is too well known for that. 1 shall discuss only a few points in connection with it which a reading of the Report, however carefully done, would hardly reveal. If in considering the official report of the journey of 1869 I am compelled to suggest certain criticisms as to its lack of accuracy, they will be given, I hope, in a spirit of fairness and truth towards all concerned.

In the early summer of 1867, Professor Powell, with a small party of naturalists and students, went into Middle Park for study. This was possibly the first class of students taken by a college professor into the West for field work, the initiation of a practice common enough in later years. He brought letters to one John C. Sumner, who was running a trading post at Hot Springs, Colorado. At that time, Major Powell (then 33 years of age) had not made any plans for exploring the Colorado River. Judging from what he himself says, it had not even entered his mind. In the preface of his Report of 1875 he writes, "The result of the summer's study was to kindle a desire to explore the Canyons of the Grand, Green, and Colorado Rivers." It has been stated by another that the first suggestions of the exploration of the Grand Canyon and the whole of the River by boat came to Major Powell from Jack Sumner, as they sat around their camp fire in Middle Park in 1867.

That fall Major Powell went back East, expecting to return in 1868 for the journey down the river. For some reason, probably because of a lack of funds, he changed his plans, and when he returned in the spring he again brought with him a large party of students for study in the field. The summer was spent in Middle Park, on the Grand, and over the Park Range. The winter of '68–'69 he passed in camp on White River, and in excursions to the Grand, the Green, and the Yampa Rivers, and around the Uinta Mountains.

Early in the spring of 1869, Major Powell went East again, and had built in Chicago four boats according to specifications made by Jack Sumner at their winter camp. About the middle of May he returned with the boats to Green River Station, in Wyoming, on the then newly-opened Union Pacific Railroad, ready for the journey down the Green and the Colorado rivers. At the railroad were gathered together the little band, frontiersmen all, with one exception, who were to accompany him. The "Emma Dean," in which the Major rode, was manned by Jack C. Sumner and William H. Dunn; "Kitty Clyde's Sister" by G. Y. Bradley and Walter H. Powell, the Major's brother; the "No Name" by O. G. Howland, Seneca Howland, and Frank Goodman; and last, the "Maid of the Canyon" by William R. Hawkins and Andrew Hall.

The expedition started from Green River, Wyoming, May 24th, 1869. Goodman left the party July 5th, at the Uinta River. The Howland brothers and William H. Dunn left the party August 28th, some thirty miles above the end of the Grand Canyon, climbed out, and were killed by the Indians. The remaining six completed the journey through the Canyon, reaching the mouth of the Virgin River on August 30th. From here Major Powell and his brother went out to Salt Lake, and Hawkins and Bradley continued by boat to Fort Yuma. Jack Sumner and Andrew Hall went by boat all the way to the mouth of the river at the Gulf of California, thus earning for themselves the distinction of being the first to make the entire trip from Green River, Wyoming, through all the canyons of the Green and Colorado rivers to the Gulf. It would be entirely out of place to attempt to relate the experiences and incidents of that memorable journey. The original report must be read to be appreciated, even though it may not be fully understood in the meaning.

When I first became acquainted with Major Powell's Report giving his account of that first exploration, it was to me the most fascinating story I had ever read. Even after completing the railway survey and finding many of his descriptions of the conditions of the canyon and the river, to say the least, misleading, I found the narrative of the adventures of the party as beautiful and as fascinating as ever. As was remarked by Mr. Arthur M. Wellington, writing in the *Engineering News* of September 21, 1889, "Major Powell was a distinguished geologist and scientific man, but he was not an engineer. He was going through a marvelous and unexplored canyon, at great personal risk, and under circumstances calculated, if any ever were, to appal the imagination and confound the judgment. He was not thinking of any such wildly visionary notion (as it must then have appeared) as the construction of a railway through the canyon, and was only anxious to produce in words the impressions of awful majesty which the gorge had produced upon him, and was likely to produce upon any intelligent observer. It is no discredit to him, if, in his effort to do this some of his statements prove to be, in the literal sense, exaggerated, when subjected to the cold scrutiny of transit and camera."

With all this in mind, however, I nevertheless experienced one of the greatest regrets of my life when I learned in later years from such good authorities as furnished me the information that a large part of the story credited to the exploration party of 1869 was taken from the experiences and notes of the expedition of 1871 and 1872. I first met Jack Sumner, Major Powell's right hand man on the river in 1869, on Dec. 13, 1889. Knowing Sumner was with Powell in his whole trip of '69, I asked him many questions about the river below, particularly about the cataract where the three men left the party and were killed by Indians, telling him that was the only rapid, and the one place on the whole river, judging from Major Powell's account, which I feared. He encouraged me in every way and assured me there were no insurmountable difficulties at that point. When I insisted that Major Powell had described the place where the three men had left him as being extremely hazardous, Sumner made no attempt to conceal his disgust. There was a trace of bitterness in his voice as he retorted, "There's lots in that book besides the truth," and turned away. At the time I was at a loss to understand Sumner's remark and the reason for his evident resentment toward the Major. Several years later I learned the fuller import of his comment on Powell and his book.

In October, 1893, I met Major Powell's brother-in-law, Professor A. H. Thompson, of the U. S. Geological Survey in Los Angeles, California. From him I first learned the true facts about the second Powell expedition. And he it was who first told me that many of the incidents and descriptions credited to the trip of '69 happened in '71 and '72, and came from his, Thompson's, own diary of the work of those years, and complained that Powell never would allow

(Photograph by Bachrach & Brother, Washington, 1901)

MAJOR JOHN WESLEY POWELL
(1834–1902)

the recognition that was rightfully due his men. Lastly, Frederick S. Dellenbaugh, a member of Major Powell's '71 and '72 expedition, says in the preface to his book, *The Romance of the Colorado River:* "When his report to Congress was published, Major Powell, perhaps for the sake of dramatic unity, concluded to omit mention of the personnel of the second expedition, awarding all credit for all that was accomplished to the men of his first wonderful voyage of 1869." And he might have added there, as he does in the text of his book, as can be seen if comparisons are made of the two accounts, how the Major gave, with dates to the very hour of the day in 1869, facts and incidents which never happened until two or three years afterwards.

It may seem strange when I state that it was not until three years after I had completed my trip through the canyons that I first knew the true facts of the second journey by Major Powell on the Colorado River. While in Utah I had heard statements in regard to it and also heard them denied, and Major Powell had spoken of it to me at one time, but in a way I could not understand. I had searched every report I could obtain and found no reference to any work up the river in 1871 or 1872. With the land survey, of course, I was familiar, but the fact is that nowhere in Major Powell's official reports or other books does he mention the second expedition, except in his short progress reports (which I did not find at that time) addressed to the Secretary of the Smithsonian Institution and dated at Washington, March 25th, 1872, Jan. 17th, 1873, and April 30th, 1874. Since that time, except for one short letter, I have never been able to find any description by Major Powell of his second journey in any written form. And yet that second journey was as notable as the first, as far as it went. It was a real exploration, with exciting adventures and with but one difference: there was no longer any fear of any great falls ahead.

In considering a few special points in the history of Major Powell's explorations of the Colorado in 1869, '71, and '72, one must not forget that the famous report of these explorations (or, as it was supposed to be, of the '69 journey only) was not only not published until 1875 (after quite accurate surveys had been made of a good part of the river), but that it was written by an eminent scholar and scientist, a former professor of a university, a trained geologist and ethnologist, a philosopher, and later Director of the U. S. Geological Survey. He was making an official report to the Smithsonian Institution and to the Congress of the United States upon a great scientific exploration. One certainly would have the right to expect in it accuracy as to facts and occurrences of the journey and the correct dates when these occurred, as well as a proper regard for accuracy in statements relating to the physical condition of the canyon walls. Considering the source and the subject,

MAJOR POWELL AS HISTORIAN 107

neither "poetic license" nor "dramatic unity" can justly be put forward in an explanation of omissions, inaccuracies, and errors.

I first read the document as I would the Gospel of St. John, with an almost worshipful reverence. Major Powell's position in the scientific world, his great service in the U. S. Geological Survey, and the Bureau of Ethnology, and his works on geology, ethnology, and philosophy place him far above any praise or criticism upon those subjects from an ordinary engineer. Yet the very position he occupied gives one, I think, the right to consider candidly and to analyze his official reports. But whatever may be the outcome such an analysis does not affect by one particle the praise due the great explorer as recorded in the beginning of this chapter. However, it is only the real facts in which we are interested at this time. Let us examine for a few moments some of the statements relating to the Colorado River and its canyons found in the official report.

First, in the report of the journey of 1869, all mention of there having been a second expedition is omitted, and in it are found incidents and accidents that happened in 1871 and '72, though they are given as occurring in 1869, even to the day of the month, and hour of the day. This was Professor Thompson's statement to me, and it is shown by the narratives written by Dellenbaugh. Thompson gave me a reason for this. Nevertheless, Major Powell stated to me personally and positively that this report of his first trip was only his diary, *"written on the spot."* Why was such a misleading account written? Dellenbaugh attributes it to a desire for "dramatic unity," Thompson to a lack of copy. I think the true reason lies in another direction. Major Powell had made the first journey through the canyons. Out of this grew that feeling of proprietorship to which I have already referred. Out of this sense of proprietorship grew the feeling, as expressed by Lieutenant Ives and many others since, that no one else would ever do the same thing again. And lastly, it would almost seem, there came a desire, as to Ives, to give the impression that no one, not even he himself, had ever been able to duplicate that first proud achievement.

Hence the suppression of the facts. Powell undoubtedly assumed that no one of his men would ever publish an account of the two expeditions. He counted rightly on even so distinguished a man as Professor Thompson. Several times after 1893 I tried my best to persuade Thompson to write and publish the story of the second journey. While he would not advance a reason for his decision, he not only refused to write it himself, but he always declined to furnish me the data for that purpose. Major Powell went even farther than Lieutenant Ives. He suppressed the facts, and he also made positive statements on the lecture platform and clear implications in his writing that no other successful trip had been made down the river since his first one, even part of the way. To his dying day he would not give any one else much credit, not even his own men. He seemed to feel worse than did Ives, who had stated that if Johnson went up the river *before* him it would take away all the glory from his expedition. Major Powell felt (at least, he acted that way) that if any one *followed* him his own glory would be lessened.

Frank M. Brown, who talked with the Major in 1889 on the proposed railroad survey, told me afterwards that the whole conversation clearly revealed to him Powell's obvious resentment of the idea that any one else might put foot in *his* canyon or on *his* river. My own impressions of Major Powell's feelings in this matter were exactly those of President Brown. I never met Major Powell personally until 1892. I then called upon him at his office at the Geological Survey. The following account of my visit I take from my note book, just as it was then written when I returned to my hotel that afternoon:

Washington, D. C., May 16th, '92.

Called on Maj. J. W. Powell this afternoon, 1 o'clock. He received me very cordially. He is a very nervous man.

I first asked him for a letter to Professor Langley of the Smithsonian Institution. He gave it very cheerfully.

Then, simply to open the conversation, I remarked that I had seen one of his old friends not long ago— Jack Sumner. He answered in a perfectly indifferent voice and manner, "Is he alive yet?" Only that and nothing more.

(This Jack Sumner is the one man who made Major Powell's first journey a success—so everyone says ——and Sumner feels sore towards Powell. He complained to me of Powell's treatment.)

We then talked on some time about the canyons of the Colorado. Two or three things struck me.

First. Major Powell's whole manner and conversation impressed me with the idea that he felt I had been trespassing on his *private property,* on the Colorado.

Second. He seemed to *labor* to impress me with the fact that his journey was entirely different from mine —to glorify his work and belittle mine. Not one word did he express of compliment at my final success, but rather sneered at any value in my work. Major Powell made such remarks as these:

"I went through without knowing what was before me. You knew all about it beforehand, and had a map, as I had when I went through the second time."

(In all my lectures and writings, I have always given this credit and praise to him; but it sounded very queer coming from him to me, in the *petulant* manner in which he expressed it.)

He also said, "I was in the Grand Canyon all winter."

Major Powell asked, "What would a railroad through the Canyon do?" Then, seeming to notice or realize the snappish manner in which he had asked it, said, "I'm asking for information. I know nothing about such things. Never studied the subject."

I explained some of the business. He said: "There is a great deal of minerals, especially silver, in the

Canyons. I examined the whole field last fall."

Major then struck off on a new line, and labored to explain his account of his first Colorado River trip. Said it was only his "diary written on the spot." He had "never published anything else," and that, "its statements were not in the least exaggerated."

I simply listened. Let him do all the talking as far as possible. The whole interview was a most extraordinary one to me—that is, coming from a man in his exalted position to me, an ordinary, everyday railroad engineer.

At the Irrigation Congress held in Los Angeles in 1893, by special request Major Powell delivered an address upon the Grand Canyon. He very properly glorified his first trip, but he made no mention of any other, either his second trip or my successful journey of 1889 and '90, but made the positive statement that no one since his first trip had passed by boat through the canyons. After the lecture I pushed my way up to the platform and offering my hand said: "I simply want to let you know, Major, that I was in the audience and heard you." The next evening, at the urgent request of some of my friends, I addressed the Congress on the same subject, giving a few facts as they really were.

Major Powell had some special reason for not wishing to acknowledge that there had been a second journey down the river, not even his own. As has already been stated, I never knew the real history of his second expedition until after I had completed my own river journey. In the spring of '91 a curious mention of it came to my ears. I was invited to hear a paper read by a Mr. Walter H. Graves, a former member of the U. S. Geological Survey, on his trip through the canyons of the Colorado. I very naturally accepted the invitation, and was very much surprised at what I heard. Never having read any such statements in any of Major Powell's reports, I concluded to go to headquarters and have the statements of the lecturer confirmed or denied, and also to learn from Major Powell himself if any second expedition had ever been down the river and through all the canyons in boats.

On March 10, 1891 I wrote to Major Powell, asking him about the 1871–72 expedition of which Mr. Graves had spoken. Major Powell's reply was studiously evasive. I put the question to him direct, but though he spoke of the river work of 1869 and the land survey of 1870, he made no slightest mention of his having made any second river journey. His letter denied effectually enough the claim of Graves to having made a river voyage, but it sought to perpetuate the strange secrecy in which he chose to keep the fact of his own second Colorado River expedition.

Answering my first question as to an actual survey of the river itself, Major Powell claims all the work of survey by boats and navigation of the river for his 1869 expedition, just as he did in his Report of 1875,

(Photo by Clinedinst, Washington, 1902)

ALMON HARRIS THOMPSON

Major Powell's brother-in-law and his chief assistant in the second Powell expedition, 1872

MAJOR POWELL AS HISTORIAN 113

denying by silence the work of Prof. Thompson and the other men in 1871 and 1872.

To show that the same desire to suppress the facts existed almost to his dying day, I submit the following from the preface to Dellenbaugh's book, *The Romance of the Colorado River*.

"Washington, January 6th, 1902.

Dear Dellenbaugh:

I am pleased to hear that you are engaged in writing a book on The Colorado Canyon. I hope you will put on record the second trip, and the gentlemen who were members of that expedition. No trip has been made since that time, though many have tried to follow us. One party, that headed by Mr. Stanton, went through the Grand Canyon on its second attempt, but many persons have lost their lives in attempting to follow us through the whole length of the Canyons. I shall be glad to write a short introduction to your book.

Yours cordially,
(Signed) J. W. POWELL."

Since it comes from such a source, one has a right to analyze such a letter. It is a fair sample of the Major's loose and inaccurate writings on the history of the Colorado River, and fairly open to criticism.

First, it is gratifying to some of us to know that, after suppressing for thirty years the fact of his second trip and using its experiences to embellish the account of his first expedition, the Major is pleased to learn that Dellenbaugh is about to tell the facts in the case—but that is neither here nor there.

Second, the statement, "No other trip has been made since that time, though many have tried to follow us," contains two positive misstatements. It has been suggested that this assertion should be taken with the next sentence, and means "to follow us through the whole length of the canyons," that is, from Green River, Wyoming, to the Virgin. This explanation does not suffice for this reason: Leaving out for the present all considerations of the railroad survey of 1889 and '90, two other parties had already started from Green River, Wyoming, and had successfully gone not only "through the whole length of the canyons," but had traveled by boat on the Colorado, hundreds of miles farther than the Major or any of his men (except Hawkins, Bradley, Sumner and Hall) had ever traveled. These expeditions were that of George F. Flavell in 1896 and that of Nat. Galloway in 1897. These facts had been published in the newspapers of the day, and an account of one of them had been furnished directly to officers of the Geological Survey in person, in Washington by my first assistant, John Hislop. Ignorance of these journeys, if that should be pleaded, is no better justification for such a person's writing such a letter on such a subject than ignorance of the law would be in a murder trial.

And again, the statement that "though many have

MAJOR POWELL AS HISTORIAN 115

tried to follow us" (meaning, of course, *and failed*) is also wrong. No one really attempted "to follow through the whole length of the canyons," if that is what was meant, except the two successful parties of Flavell and Galloway. It is barely possible that reference here was intended to H. M. Hook, whose grave was found on Green River in 1871. Hook, however, was not attempting to follow the Major. He was only going to certain supposed mines on Green River, and when warned of the danger in the canyons is reported to have said that if Major Powell could go through, he could. But this had reference not to "the whole length of the canyons" but only to the upper Green River canyons as far as the mines.

Third, "one party, that headed by Mr. Stanton, went through the Grand Canyon on its second attempt." This is in exact accord with the interview I had with Major Powell soon after completing my journey—an attempt to belittle the work of the expedition of 1889 and '90, by saying "on its second attempt," and restricting its extent to the Grand Canyon alone, whereas in fact we passed through two of the canyons of the Green River, below Green River Station, Utah, and all of the canyons of the Colorado to the Gulf of California. But this, I suppose, was considered no trip on the Colorado, seeing that it did not start at Green River Station in Wyoming!

Fourth, and most surprising of all, is this: "but many persons have lost their lives in attempting to follow us through the whole length of the Canyons." Whatever that may mean, it is entirely misleading. The reference may be to Hook. Still, he was but an individual who lost his life. There were no others. The Major may have had in mind President Brown and the two boatmen of his party who were drowned in 1889 in Marble Canyon. In that case, however, he is driven to the other horn of the dilemma, and is thus forced to admit that there *was* another trip. Yet, although three men were drowned, the expedition was successfully carried through to the end. Hence Powell's whole letter to me is shown to be a matter of special pleading and apparently, intentional suppression and misrepresentation of the facts.

It has always been supposed by some of us that naturally Major Powell had kept in the field a detailed record of his memorable journey down the Green and Colorado rivers in 1869. This was the opinion of F. S. Dellenbaugh of the second Powell expedition of 1871 and '72. Though he greatly desired to see the record, up to the time of Major Powell's death in 1902 Dellenbaugh felt a delicacy in asking the Major for it. Subsequently Dellenbaugh made an attempt to find the journal, but with no success.

In February 1907 I went to Washington and began my search for Major Powell's field notes or diary, and also the diary kept on the same expedition by Jack Sumner, which he had loaned Major Powell and for which I had an order on the Geological Survey. I first called on Dr. Charles D. Walcott, who later succeeded Powell as Director of the Geological Survey but who was then Superintendent of the Smithsonian Institution. Dr. Walcott knew nothing of any such papers in the records of the Survey. He suggested that Prof. G. K. Gilbert, of the University of California, to whom all of Major Powell's papers had been left for disposition after his death, might have or know something of them.

I wrote to Prof. Gilbert and soon after received a reply stating that he and Mrs. Powell had gone over all of Major Powell's papers, but nowhere among them had they seen any diary or journal written by Sumner, the particular thing I had inquired for at that time, but that he had ordered his clerk in Washington to make a search for the papers. The same evening (February 25th) I called on Major Powell's widow and daughter. Mrs. Powell said that she and Prof. Gilbert had found among her husband's papers no diary or journal of his or Sumner's on the 1869 expedition; that the Major seldom kept a diary or notes, and that she did not believe any existed. The Major had a wonderful memory and depended almost entirely on that.

Even this did not satisfy me. I called next day on Mr. W. H. Holmes, Chief of the Bureau of American Ethnology, where Major Powell as Head of the Bureau had spent the latter years of his life. Mr. Holmes knew nothing of any such papers in the records of the Bureau, but introduced me to Miss Mary C. Clark, Major Powell's former secretary. She, too, had never seen or heard of any diary or journal or notes kept by Major Powell in 1869. She promised, however, to make a thorough search through all the papers of his office. For this I waited about two months, in the meantime bothering Miss Clark and every one I could think of with various letters of inquiry. Finally Miss Clark informed me that she had searched everywhere, through the Major's desk, book shelves, in every safe, closet, and cubbyhole in the office, that she had found nothing, and that she did not believe any such records existed.

This again did not satisfy me. I went to Washington for a final search. On April 18th, Miss Clark repeated her belief that no diary or journal of 1869 existed, for the same reasons that Mrs. Powell had given. But I continued my inquiries. By this time the patience and interest of the whole office force—and I confess my persistency and hopes—were fast fading away. Then Miss Clark remarked: "By the way, over there in the corner under that old desk, which has not been used for years, are two or three bundles of old papers. I haven't the least idea what they contain, but if you wish to, you may look them over."

The dust and cobwebs of years did not stop me. I took off my coat, rolled up my sleeves, and got to work with a feather duster. I opened the bundles and spread out their contents on the floor of Major Powell's old office. Almost the first thing I laid my eyes upon was the long-lost, long-sought-for Journal of the Exploration of 1869, in Major Powell's own hand writing—the original itself! There were also found other journals, reports, note books, loose notes, and sketches of other early Rocky Mountain region explorations.

Upon obtaining Mr. Holmes' permission, with pen, ink, paper and a private room to myself I made a verbatim copy of the much-coveted document. After I had completed making the copy the journal was carefully wrapped up, labeled, and locked up in the safe. Later I went back to the Bureau. Mr. Holmes himself got out the journal and I made a second careful comparison and checking of my copy with the original. The only reference to this field journal that I have found in any of Major Powell's writings appears on page four of the Preface of his book, *The Canyons of the Colorado,* published in 1895, which is as follows:

"My daily journal had been kept on long and narrow strips of brown paper, which were gathered into little volumes that were bound in sole leather in camp as they were completed. After some deliberation I decided to publish this journal, with only such emendations and corrections as its hasty writing in camp necessitated."

When I found the journal, it was, as stated above, written in pencil on long narrow, yellowish-brown paper, but the sole leather bindings were gone, the whole record being tied together at the top in one batch, or pad, with a tow string.

This field journal begins under date of July 2nd, the third day after the party arrived at the mouth of the Uinta River. No journal or notes of the experiences of the expedition of 1869 from the time of leaving Green River, Wyoming, May 24th, to July 2nd, have been found. I do not think the Major kept any journal before July 2nd, and for this reason: After visiting the Uinta Agency the party again started down the river July 6th, and the camp that night is noted in the journal as "Camp No. 1." After that the camps are noted in consecutive numbers from one to forty-four. There the journal ends, except that there are no entries at all in the journal from July 7th to 19th, and the first camp noted after the latter date is "No. 11."

Besides this journal, now in the Bureau of American Ethnology, there are still preserved the Major's original geological and astronomical notes of his first expedition, written on the same common, yellowish-brown paper and bound together in the same way, papers which I also resurrected from the bundles referred to. The journal is not signed by Major Powell but is in his own hand-writing, as testified to by his former private secretary. There is no doubt that this paper is the original journal kept and written by Major Powell on the 1869 expedition. Comparison

MAJOR POWELL AS HISTORIAN   121

with the dates and incidents in the published report is absolute proof, when taken together with Major Powell's statement and Miss Clark's identification of the hand writing.

One most puzzling point had been cleared up by the discovery of this original journal. From the time in 1889 when I passed through Cataract Canyon and navigated the first two hundred and sixty miles or so of the Colorado River until I found and read the journal in 1907, it seemed to me unaccountable that a scientific man could have written such inaccurate descriptions of portions of Cataract Canyon as are found in Major Powell's official report.

The farther I went through the canyons the more my astonishment increased. Comparing the original journal with the report and instrumental data gathered since 1875, I find this fact standing out clearly: Every entry in the journal, written day by day as the expedition progressed, is absolutely accurate and true, as far as it would be possible for any man to gather such data on such a trip. Furthermore, the descriptions in the journal of the walls of the canyons, their nature, form, height, and so on, and the estimates of the fall of the river at the various rapids noted are remarkably accurate, even in detail, as is proved by photographs, surveys, and maps made since that time.

As to the journal as it exists in its original form for that portion of the trip which it covers, therefore, not one word of criticism can be offered. For its purpose it is most complete. Furthermore, even testing these original notes by the more accurate data acquired by instrumental work in the canyons since that time, one finds the journal to be accurate in every instance and its statements correct as close approximations. His field notes recording the form and condition of the canyon walls, the nature, height, and fall of the rapids, and so forth, are unquestionable.

On the other hand, the description of the physical features of the canyons, and the nature and fall of the rapids of the river, when given in the report and Powell's later writings, are in many instances distorted and exaggerated. In some instances, particularly with regard to the fall of rapids, the estimates of the journal have been doubled and almost trebled. Of the point of steepest fall of the river in Cataract Canyon, the journal says: "Three long portages, when the river fell, by estimate, 42 feet." Of this same spot the report states: "the distance being less than three-quarters of a mile, with a fall of 75 feet." By our railroad survey at this spot, the steepest in that canyon, the greatest fall is shown to be *55 feet in two miles,* though the heaviest fall for a distance of about a half-mile approaches Major Powell's journal estimate of 42 feet. Nowhere in Cataract Canyon did we find a fall of 75 feet in three-quarters of a mile, nor was any such fall recorded in the original journal.

Of the noted "Sockdologer" rapid, the journal of '69 says: "chute one-half mile long, fall 30 feet, probably." The official report states: "There is a descent of perhaps 75 or 80 feet in a third of a mile." Of this same rapid, Jack Sumner, in his diary written during the trip of '69, records: "A long rapid, with fall of 30 feet," showing that there was no disagreement at the time as to the probable fall of that rapid. The "Sockdologer" has an actual fall of about 35 feet. In this instance Major Powell's original estimate was somewhat too low, but in the report, for some unknown reason, he unaccountably multiplied his first figures nearly three times.

There being no journal or diary from May 24th to July 2nd, the report published in 1875 for that part of the journey was written from memory, and, as Professor Thompson says, from his diary of the trip of 1871. I have never seen Professor Thompson's diary, but with the publication of the complete record of the Second Powell Expedition by Dellenbaugh (*A Canyon Voyage, 1908*) we have the opportunity to verify Professor A. H. Thompson's statement as regards the origin of a part at least of the report of the first exploration. Dellenbaugh states in his preface, "The material collected by this expedition was utilized in preparing the well-known report by Major Powell, the second party having continued the work inaugurated by the first and enlarged upon it, but receiving no credit in that or any other government publication."

By comparing the report with Dellenbaugh's narrative on that part of the Green River above the Uinta and from other information I have gathered, one concludes that the greater part of the story of '69, exactly as Prof. Thompson told me, was written from the notes made in '71. Of course, the most notable incidents, such as the loss of the "No Name," the finding of the record of Ashley, and so forth, did occur in '69. Furthermore, a few points, such as Flaming Gorge, Red Canyon and Lodore Canyon, were named in that year. But the naming of almost all the others and the incidents connected with their naming did not come about until 1871, for instance, these which were credited to the trip of 1869, and so recorded afterwards in the report with the day and the month as follows:

May 30th, "We name it Horseshoe Canyon" and "so we adopt the names Kingfisher Creek, Kingfisher Park, and Kingfisher Canyon."
June 4th, "We call this Swallow Canyon."
" *8th,* "So it occurs to me to designate this part of the wall the Cliff of the Harp."
" *12th,* "We adopt the name Disaster Falls."
" *22nd,* "So I name it Island Park."
" *24th,* "So we name it Split Mountain Canyon."

The true facts are as follows: "Swallow Canyon" was named by Dellenbaugh. The occurrences leading to the naming of the "Cliff of the Harp" are specially noted and commented upon as happening in 1871 by Dellenbaugh on page 43 of his *A Canyon Voyage.* Even the name Disaster Falls was adopted and so recorded by the topographers in 1871. Of "Island Park" Dellenbaugh says, "The Major was for a time [1871] uncertain whether to call this 'Rainbow' or 'Island Park,' the decision finally being given to the latter." The name Split Mountain Canyon came from the peculiar geological formation of the mountain, and this was not known until 1871. The description by Major Powell of the climb as recorded on July 24th, '69, undoubtedly did not occur until about July 4th, 1871. The name 'Split Mountain' was not even adopted at that time. Dellenbaugh says (page 57), "We then called this 'Craggy Canyon,' but later is was changed to 'Split Mountain.' "

Comparison of the Report with the original journal from July 2nd to August 28th clearly shows that the journal lay beside the Major when he wrote the report, and that he consulted it as he wrote, but in a very careless manner. Further, for some unexplained reason, the report in many instances is an exaggerated statement of the journal notes, and in one at least an absolutely contrary statement.

In the report are the following:

July 15th, '69. "And we name this Tim-Alcove Bend."
Same date. "So we name it Bow Knot Bend."
July 17th, '69. "We name this Bonita Bend."

All of these names were first given in 1871. The last one was selected and given by E. O. Beaman. I am also informed that many of the detailed experiences of running the rapids and climbing the canyon walls in this part of the report were the actual experiences of 1871 as recorded in the diaries of that year, with the names of the men of '69 inserted to make the deception more plausible.

Where both the journal and report exist from July 20th to August 28th, the inattention and exaggeration are clearly marked. The deadly parallel shows this only too plainly:

From the Journal	*From the Report*
July 20th. "Climb 'Cave Cliff' with Bradley" (following which is a short description of caves, pools, etc.).	"This morning Captain Powell and I go out to climb the west wall of the Canyon." (An amplified description of country follows.)
July 24th. "Only made ¾ mile today. Three long portages when river *fell, by estimate* 42 feet. Huge rocks across the river."	"We examine the rapids below. Large rocks have fallen from the walls. . . ." "We are compelled to make three portages in succession, the *distance being less than* ¾ *of a mile with a fall of seventy-five feet.*"

This fall, we are to believe, grew from 42 to 75 feet since the journal was written on the spot! The truth is there is no 75 feet fall in three-quarters of a mile anywhere in Cataract Canyon. The greatest fall, as shown by the railroad survey, the leveling being done with a Y-level, is 55 feet in two miles, so that perhaps the original estimate of 42 feet was somewhat nearer correct. The 75-foot fall must simply be attributed to Powell's imagination.

<table>
<tr><td>*Journal*</td><td>*Report*</td></tr>
<tr><td>*July 28th.* Made two portages today, one very long one at noon. After dinner ran a long chute about half mile, very narrow, very rapid down the slope of the rocks. It had a marble floor. *Then the Canyon was rapid, narrow, straight,* the walls rising from the water's edge, *and running back with two grand steps that gradually came down to the water's edge. Between these smaller interrupted steps.*</td><td>"We made two portages this morning. One of them very long. During the afternoon we ran a chute more than half a mile in length, narrow and rapid. This chute has a floor of marble." . . . Describes running chute. *"After this the walls suddenly close in,* so that the Canyon is narrower than we have ever known it. The water fills it from wall to wall, giving us no landing place at the foot of the cliff; the river is very swift, *the Canyon is very tortuous* so that we can see but a few hundred yards ahead; *the*</td></tr>
</table>

Report (continued)
walls tower over us, often over-hanging so as to almost shut out the light."

What a wonderful change came over the canyon and its walls between 1869, when the journal was written and 1874 when the report was written! From a "straight, narrow canyon," it drew itself up into "*a canyon very torturous so that we can see but a few hundred yards ahead,*" The "two grand steps that gradually come down to the water's edge" and the smaller steps had entirely vanished by 1874, being replaced by canyon walls which "tower over us, often overhanging so as to almost shut out the light."

August 14th. "Made a portage from camp, then ran two miles to bad falls in narrow chute, no talus, no foothold of any kind, must run it or abandon the enterprise. Good luck! Little boat fills with water twice. Chute ½ mile long. *Fall 30 feet probably, huge waves.*"

"The river enters the granite" etc. "About eleven o'clock we hear a great roar ahead, and approach it very cautiously. . . . . . *There is a descent of perhaps 75 or 80 feet in a third of a mile,* and the rushing waters break into great waves on the rocks," etc.

This is the noted "Sockdologer" rapid, and here again for some unaccountable reason the fall has increased from probably 30 feet to "perhaps 75 or 80 feet."

Since everyone while on the river, including Major Powell himself, estimated the fall to be from 30 to 35 feet, it seems clear to me that Major Powell put it in his printed report five or six years after (with his original diary by his side), as "perhaps 75 or 80 feet," not, as Dellenbaugh would say, for the purpose of dramatic unity, but for poetic exaggeration. In his first published account of his exploration of '69 (republished in 1870, in London, in Bell's *New Tracks in North America,* Second Edition, page 562) Major Powell writes of certain cataracts between the Little Colorado and the mouth of the Virgin as "often falling at a plunge from 5 to 20 feet—the greatest being 22 feet." But in the report of 1875 one of these same cataracts is stated to have had a fall of "75 or 80 feet"—quite a difference, to be sure.

In a number of points this account differs from both the journal and the report. In this article written in 1869 is given the statement with regard to the fall of the rapids already noted. The following quotations show the difference between the Report published in 1875, and the article of 1869.

From the Report	*From the Article*
July 24th, '69 "Make three portages in succession, the distance being less	"Rowed out into Cataract Canyon . . . never finding any fall greater

From the Report (continued)	*From the Article* (continued)
than three-fourths of a mile, with a fall of seventy-five feet." *August 3rd, '69* After describing different features of the long Canyon from Dirty Devil River to Paria Creek, the Report asks: "From which of these features shall we select a name?" and answers: "We decide to call it Glen Canyon."	than nineteen feet in this canyon. "On the 31st of July we reached the mouth of the San Juan at the foot of Mound Canyon." Describing the next canyon: "The little valley of the Paria River terminates this Canyon. . . . We named it Monument Canyon."

It would have been perfectly proper in 1871 or at any time afterwards for him to change the names of any or all of the canyons from these given them in 1869 to any other names he saw fit. But when in his report he states positively that the names there set forth were selected in 1869, it seems to me that Major Powell's original diary and his *published* article of 1869 contradict him flatly. I have no comment to make. The Major's own words are all that need be given.

Another and to me the most surprising and unnecessary instance of the suppression of facts is to be found in Major Powell's not mentioning in his report or in any of his writings, lectures, or conversations that he had worn a life preserver on his first expedition on the river. The fact that he wore a life preserver was so carefully concealed that while the members of the 1871–72 party knew that he wore one on that second journey, none of them, not even his brother-in-law, knew that he had one in 1869.

Dellenbaugh records the use of the life preserver in '71 and '72 and how it several times saved Major Powell's life, but states positively that "on his first trip he did not have any." Dellenbaugh made this statement because he had never heard of the life-preserver, and because he had spent several years in personal association with Major Powell on the river and elsewhere, and while the Major talked hundreds of times on the journey of '69, he never once mentioned his life preserver and how many times it had saved his life on that first trip.

Here are the facts as related to me first by Jack Sumner years ago. When Major Powell came out from Chicago with his boats in 1869 among other supplies for the expedition he brought one rubber life preserver for himself, and he wore it through the whole expedition to near the end of the Grand Canyon. Why he did not bring one for each of his men I do not know, except that he had no conception of the dangers of the rapids, and brought one for himself, a one-armed man. This was perfectly proper, for a one-armed man would live but a minute in any of the rapids of that foaming river. He therefore did the proper thing in providing a safeguard for his life, but why did he studiously suppress the fact for thirty-three years? How could he for so long a time talk of every other incident of the journey and close his mouth tightly even among his personal friends when the words "life preserver" came to his mind? Did he feel that it would take away from the glory and heroism of that first trip, as would acknowledging the second or any other trip that came after? If so, what a mistaken notion! From the time Major Powell lost his arm at Shiloh in heroic defense of his country to the day of his death, no one ever doubted his personal, physical bravery. Unfortunately it took another kind of courage to tell the whole story of the first exploration of the Colorado.

On pages 48 and 49 of his report the following is recorded under date of July 11th, 1869: "When another wave rolls our boat over, I am thrown some distance into the water. I soon find swimming is very easy and I cannot sink." And further: "Our rolls of blankets . . . were thrown out . . . but I succeeded in catching one of the rolls of blankets as it drifted by when we were swimming to shore; the other two were lost." A one-armed man towing a roll of blankets to shore in a rapid of the Colorado was more than I could understand until the following correspondence was had:

"New York, February 19th, 1907."

Mr. J. C. Sumner,
    Vernal, Uinta Co., Utah.

My dear Friend Jack,

. . . .

The most important point, however, which I want you to settle for me, and to do it at once, is this: Did, or did not, Major Powell have a life preserver on his first expedition, and did he wear it on July 11th? As recorded on page 48 of his first report where he says, "I succeeded in catching one of the rolls of blankets as it drifted by when we were swimming to the shore." Now, it has always seemed to me that it is utterly impossible for a one-armed man in a rapid, such as you know on that river, to do those two things unless he had a life preserver on, and I have always been under the impression that you told me that the Major had a life preserver. If I am correct in this please answer me, and whether or not on all the rough water, during your expedition, he wore it. And, again, did any of the rest of the men have life preservers at that time, that is, in 1869?

. . . .

(Signed) ROBT. B. STANTON."

"Vernal, Utah, February 26, 1907."

Dear Friend Stanton:
    Yours of the 19th received.

. . . .

134   COLORADO RIVER CONTROVERSIES

Yes, Major J. W. Powell had a life preserver made of rubber to be inflated when needed. It was the only one in the outfit, and as Powell was in one sense under my care, I made it my special duty every morning before starting to inflate it and lay it by his side on the stern seat of the "Emma Dean." He certainly wore it in the rapids you mention, and in every other one that looked dangerous to me, as I would not have run into a bad rapid hampered by an unnecessary dead weight, and he certainly would have been drowned in any bad rapid without a life preserver. Yes, I think there is one left besides myself of the old party—Wm. R. Hawkins, Eden, Graham Co., Arizona.

. . . .

(Signed) JACK SUMNER."

New York, March 4th, 1907.
Mr. William R. Hawkins,
    Eden, Graham Co., Arizona.
My Dear Mr. Hawkins:

. . . .

Over one point there has been a good deal of discussion in years past and Sumner has certainly settled the matter finally; that is, whether Major Powell wore a life preserver on his first trip through the river, and whether it was that fact of having on a life preserver that saved his life several times when he was thrown into the rapids. Sumner writes me that this is a fact, described the life preserver, and says that he himself inflated it every morning ready for the

Major's use, and that the Major wore it in all of the bad rapids.

I wish you would, in a short note, confirm these facts, if they are entirely correct.

. . . .

(Signed) ROBERT B. STANTON."

"Eden, Arizona, March 11th, '07.
Mr. R. B. Stanton,
My dear Sir:

. . . .

I have now by me, and in front of me on my desk, the very life preserver that Powell wore in all bad places on that trip. Of course it is dried up and the rubber cracked. It was given to me by Major Powell at the time the boys left us.

. . . .

(Signed) WM. R. HAWKINS."

I finally induced Hawkins to present the life preserver to the Smithsonian Institution for safe-keeping as a relic of the first exploration of the Grand Canyon. He first sent it to me, and it was in my possession for some time. I sent it on to Washington.

The question as to whether Major Powell wore a life preserver in 1869 may in itself seem of little importance, but its bearing upon his whole report is significant. The suppression of certain facts and the insertion of other facts with the year, day, and hour of their happening in 1869 which did not happen until two years after cast strong suspicion upon the whole document.

That there was a distortion of the facts and conditions which cannot be accounted for by sentiment, poetic license, or "dramatic unity" is shown also in the character of many of the illustrations used in the story of the '69 journey. No photographic instruments were taken on the first exploration. In '71 and '72 a photographer and instruments were enclosed. Many of the photographs taken then were used to illustrate the trip of '69—without explanation, although the report was not printed until 1875. Rather an amusing want of care and oversight is shown in some of the illustrations, that is, the changes made in the pictures were not sufficient to cover up their identity. Under date of "May 24th, 1869," figure 4, "The start from Green River Station," shows *all ten men in three boats,* whereas, in '69 there were four boats, and the picture given is from the photograph of the three boats that were used in '71. Figures 6 and 18 show in the foreground cameras which were there only in '71 and '72, and the arm-chair on the boat which was used by Major Powell. That arm-chair did not exist in 1869.

Most absurd and deceptive are the purely imaginary frontispieces, figure 1, and the ridiculous picture, figure 10, of the "Wreck at Disaster Falls." Why was it necessary to invent such a picture when there existed at that time a photograph of the exact spot, taken in '71, showing the rapid and the walls of Lodore Canyon as they really are, into which very appropriately might have been put the boat and the drowning men?

All the foregoing has been presented with the purpose of proving conclusively that, with due respect to Major Powell as a scientist and as the first real explorer of the Colorado River, his now famous Report of 1875 is demonstrably inaccurate and, it would seem, deliberately misleading on a number of counts. Having been shown, then, that the Major was undoubtedly guilty of suppression of the truth and unblushing exaggeration in these various matters of lesser import, the unprejudiced reader cannot, I trust, doubt the conclusions which I draw after a consideration of the Major's and other accounts of a much more important and regrettable incident of the voyage of '69. I refer, of course, to the Separation Rapids affair.

But before turning to the Major's version let us look at the narratives of two members of the 1869 expedition, the accounts of J. C. (Jack) Sumner and William Rhodes Hawkins.

CHAPTER II

WILLIAM HAWKINS' STORY

As already related, in hunting up some evidence on the Powell exploration of 1869, I became acquainted by letter (I never met him) with William R. Hawkins, the cook of the expedition. Hawkins was the youngest man of the party. At my request, the following sketch was sent to me in the summer of 1907. The exploration of 1869 was of such importance that any light cast upon it, especially from the inside, even from a subordinate member of the party, is of peculiar interest.

One of the results of the wild life of the early frontiersmen, even of those of little education, was the sharpening of their wits and the clearing of their vision so that they saw things in their true light and gave them their true meaning. They thought and spoke with no conventional reason for hiding the truth, that is, unless they were selling a tenderfoot a mine. Hawkins' discussion of the traits of Major Powell's character is sincere, and just such an estimate as would be expected from a free trapper of the West.

The question of fact and fancy in the stories of Hawkins, Sumner, and the Major is discussed else-

WILLIAM HAWKINS' STORY

where. The statement made by Hawkins that at one time on that memorable 28th of August, '69, the decision had been reached, however reluctantly, for the whole party to leave the river and abandon the expedition for a time has come to me from sources other than this narrative by a member of the party. The occurrence was talked of in after years by other members of the two expeditions, so that I am satisfied that Hawkins' story, in its broad facts if not in its minor details, is correct.

In his report Major Powell states, "At one time I *almost* conclude to leave the river." This was actually written years after the occurrence, not in Major Powell's original journal, which is still in existence. There is needed but the erasure of one word of the statement in the Report to make it agree absolutely with Hawkins' account.

Major Powell corroborates Hawkins' claim that he was the first one (or, as Powell says, other than Walter Powell) to promise to continue down the river, and after him, then Sumner, Bradley, and Hall. Hawkins' account of this particular incident may be the more nearly correct as to details, as can be seen by referring to Powell's original journal. Both were written from memory, neither man having made notes at the time of the main happenings, much less of the minor details. But even if it is the more nearly correct, that in no way reflects upon the bravery of any man in that little band of heroes. Everything seemed just then to be against a continuing of the journey, and that particularly severe rapid had its influence upon every member of the party. Even Hawkins admits that the men who left the party "had more than an even chance to get out by land."

Herewith I present Hawkins' story of the circumstances and events leading up to the Separation Rapids episode and its attendant tragedy.

In the summer of 1868, I was camped at Hot Sulphur Springs in Middle Park, Colorado, with my pack train of two mules and two horses. I was at that time trapping. My headquarters camp was 100 miles west, but in the summer I would take the furs I had caught during the winter out to Denver and sell them, and then put in three months packing provisions to my camp for the next season's trapping. I had gone out as soon as the snow was so I could cross the mountain range, and had already made two trips and was on my third and last for that summer when I heard of Major J. W. Powell's proposed expedition. Returning, I camped at Jack Sumner's place at Hot Sulphur to rest my animals three or four days. I think it was the second day after I got to Sumner's post that Major Powell's pack outfit came in and camped a short distance from where I was. It was the custom of us mountaineers as well as the Indians to find out where strangers were going and what for, since we as well as the Indians respected each other's trapping grounds. Therefore I went over to Powell's camp to find out what his intentions were.

I found the Major a very pleasant gentleman and very easy to get acquainted with. I asked him his business in that part of the country, and he stated his final object was the exploration of the Grand Canyon of the Colorado. I told him that would be an interesting trip, that I had been to some parts of it, and it was rough enough for any use, and that some of my neighbors, the Ute Indians, said the river in places ran under the ground. Powell said it was his intention to find out, and that he was acting in the interests of the Smithsonian Institution. He asked where my winter camp was. He said he intended to make his winter quarters somewhere on the White River, and asked me all about the Indians.

Sumner, Bill Dunn, and I had been talking for some time past of building us some boats and starting down through the canyons to trap. In the evening Sumner and Dunn came to my camp, and Sumner said that now we had a chance to go to Cottonwood Island down on the lower Colorado, and that he would speak to Major Powell and perhaps we might become members of his party. I told Sumner I had several hundred dollars' worth of provisions and trinkets I would have to dispose of beforehand.

Next morning Sumner and Major Powell came over to my camp, where Dunn had spent the night with me, and Sumner spoke up and said that the

Major would like us (Dunn and me) to join his party for the winter, as well as for the trip down the canyons of the Colorado. I told the Major that Sumner was thinking of selling his trading post, and that he, Dunn, and myself were going to try the canyons as far as Cottonwood Island. I told him that I had already packed my year's supplies in from Denver. The Major replied, "Those things are just what we want, also your mules and horses. I will buy them and pay you just what you can sell them for elsewhere." I told him I would think the matter over and let him know in the evening. In the meantime he and Sumner arranged a trade for Sumner's supplies at his post.

Later I went to the Major's camp and told him if I could arrange things and agree on a price for my animals, I would join his party. I showed him my bills and everything I had in stock, and he allowed me the price of the goods and for packing them over the range. He allowed me a fair price for three of my animals. For the other one we could not agree upon a price, so I kept it, as it was a favorite of mine and had carried me safely through several Indian fights. After all was counted up, outside of thirty-six traps, it amounted to $960. The traps, he said, could be used that winter, and when we got to Cottonwood Island he would replace them. He also offered me $1.50 a day for cooking for the outfit, and said that when we got through our journey down the Colorado Canyon he would pay our transportation to where we were then at Sumner's place. The Major added, "Now, as to the money you are to receive for your goods and horses, I expect to get it from the Government and also to pay your transportation back," explaining that he had money with him, and that he would let me have what I needed during the trip, and that he would be responsible to me for the remainder. He gave me a receipt for the amount due me at that time. Sumner told me that he was to receive good prices for his supplies and stores and good pay for his work, and that he would have several thousand dollars coming to him when he got through, and that he was to receive it all when we got to the end of the river journey. Later in the fall, Major Powell told me that if he could get the Government to appropriate $12,000 there would be $1,000 for each of us and $2,000 for himself for the trip.

When we finally got to the mouth of the Virgin River, the Major and his brother left us. I cannot tell how much money he paid to each one, but he gave each man some. He gave me $60 and Hall $60, and said that he would send us a government voucher for the rest, also for what he was to pay me for the provisions and animals I sold him. As we had not come to Cottonwood Island, I asked him how about my traps that he was to make good. There were thirty-six of them, and I had paid $3.00 apiece for them.

"Well," the Major said, "they got lost in Green River."

"Yes," I replied, "but you agreed to make them good when we got to Cottonwood Island."

Then he said he would allow me $2.50 for each trap and would find out what the transportation was, and send me a voucher for the whole amount. The voucher is still coming. All I ever received was $60, at the mouth of the Virgin River. Sumner is out a good deal more than I am. I saw the Major after that and asked him about it, and he told me he was still expecting an appropriation from the government. Then he would pay me.

We made our Winter camp on White River, hunting trapping, and gathering specimens. In the spring of 1869 we moved to Green River City, and waited for the arrival of Major Powell and the boats from the East. While we were camped at Green River, something happened which may be worth telling. The Major brought from the East a young man whom some friends of the Major's sent out West to see the sights and join our expedition down the unknown regions of the Great Colorado of the West. We boys noticed that this young man did not seem to like our bill of fare. I was the cook and the boys, particularly Sumner, bragged on my cooking and said I could make the best coffee in the world and a pie that would last a man a week. But it did not exactly suit the young tenderfoot. The third day, Captain Powell, who had just bought himself a new pair of boots, wanted to wash his socks. I gave him a small camp

(*Photograph taken in 1909*)
WILLIAM R. HAWKINS
Cook of the first Powell expedition of 1869

kettle and he put the socks to soak.

At noon I had the coffee in a large camp kettle, and the small one the Captain's socks were in was hid from the boys behind the large one. Then we all sat down on the ground to eat our dinner. I had done my level best that day to please our new-comer. I had dished up the coffee and they were all drinking it and feeling fine. Sumner said there was something mighty peculiar about the coffee, and asked me what was the matter with it. I took my bowie knife and stuck it in one of the Captain's socks and held it up over the kettle just back of the coffee. With the reddish brown water running off, it looked just as though I had taken it out of the coffee kettle. I yelled, "Who in hell put his socks in the coffee?" All said they had not, except the young man, who did not answer. When I asked him if he had, he said very politely that he had not, but he was getting up and leaving at the time. That was the last we ever saw of him. When the Major got his mail at the Virgin, he heard that the young man had gone home and said our grub did not suit him, and that he thought the Major had a hard crowd with him.

On the 24th of May, 1869, we started down the Green River with provisions for eight or ten months, according to Major Powell. Our provisions got wet and were lost in different ways, and finally the largest boat-load was lost in Diamond Falls on the Green River. All this reduced our rations. Though we had plenty of fish on Green River, we caught very few on the Colorado and soon found we were up against it for grub. It was not long before we began to lose our hats and clothes. I had a pair of buckskin breeches. They were so wet all the time that they kept stretching and I kept cutting off the lower ends till I had nothing left but the waist band. When this was gone I was left with a pair of pants and two shirts. I took the pants and one shirt and put them in the boat's locker, as I did not know what the law was below as to nakedness. I cut holes in my shirt tail and tied the loose ends around my legs so they would not bother me in the water. Major Powell said he was dressed when he had his life preserver on, and he always had it on when the water was bad.

I can say one thing truthfully about the Major—that no man living was ever thought more of by his men up to the time he wanted to drive Bill Dunn from the party. A description of this you will find below. I have only written here a few facts on things that happened on the Colorado expedition. There is no revenge in my heart. With the best of feelings towards the Major, I have written this because I think his Report is somewhat lacking. I am willing to do more by him than he ever did by any of us men. I am willing to call him a brave and daring leader, but I do not think the boys who left the party, the Howland brothers and Bill Dunn, under the circumstances herein mentioned, deserve to be branded as cowards. I do not wish to cast any discredit on Major Powell's Report or upon his memory of the Colorado expedition. But in justice to Dunn and the Howland brothers I must say that the account in the Report which accuses them of cowardice is entirely wrong, and that it was made to cover up the real cause of their leaving.

The trouble with the Howland boys began away back at Disaster Falls, where their boat was lost, but with Dunn it began only a few weeks before he left the party. At noon one day when the boats were being let over a bad place, Dunn was down by the water's edge with a barometer, taking the altitude. He was also assigned the post to look after the rope fastened to the boat and held by Sumner and others. By some means Dunn was thrown into the river, but he caught the rope and finally got out. In this he got wet a watch that belonged to the Major. At dinner the same day Major Powell told Dunn that he would have to pay for the watch or leave the party, which was impossible at that point. Dunn told him a bird could not get out of that place, thinking the Major was joking, but all of us were very quickly convinced that every word the Major said was meant. Dunn said he could not leave then, but that he would go as soon as he could get out. The Major then said he would have to pay one dollar a day for his board until such time as he could get out of the canyon. The rest of us sat listening as we ate our dinner. As Sumner was the oldest of our crowd, that is, the two Howlands, Dunn, Hall, and myself, we naturally looked to him as our spokesman.

Sumner told the Major he was surprised at what he had said to Dunn, and the Major said he was running the expedition. But Sumner said that was one thing he could not do—compel Dunn to leave the party or make him pay for his board. Walter Powell, better known as Captain Powell, took up the quarrel and thereby came near getting shot. We all considered the Captain demented because of his imprisonment in Andersonville prison. Had it not been for this, I doubt very much if the Captain would have made the entire trip.

When we came to the rapid where the Howland boys and Dunn decided to leave the party, we all looked at it and Hall and I had our course picked out on that rapid for the morning. The Major and some of the other boys went across the river to look to see if there was any chance to let the boats over it by ropes. On the following morning when I raised up from my bed about daybreak, I saw the Major walking up and down on a patch of sand and I called to Andy Hall, who shared the bed with me, and asked what the Major meant. He said he saw some of the boys and the Major in council the evening before, but could not find out what was the matter.

By that time the Major walked up to where I was making a fire and said, "Billy, how much flour have you now?" I said, "Very little," and showed it to him, and he told me to make it all into biscuits as near one size as possible. I asked no questions, excepting if we wanted any dried apples cooked, as we had about 100 pounds of them, and about 75 pounds of coffee. That was our entire stock of provisions. The Major said I need not cook any apples.

Andy Hall drew my attention to a group of the boys that were off to one side counciling among themselves. Bradley came up to the fire and I asked him if it was not coming to a show-down. He said he could not understand the Major, but there was something going to happen. I did not know or care what happened, as I was sure Andy and Bradley would stand by me in anything that was reasonable.

I called out to come to coffee. All came and drank the coffee and ate the bread, and there was not one word said, one to another. When we got through the Major said, "Billy, you take this bread and divide it just as equally as you can." I made nine piles and put one biscuit in each pile till my supply was gone. I don't remember how many there were to the man, but there were three or four left over and I said we would give these to the Major, but he said his share would have to do him.

The Major, the Howland brothers, Dunn, and Sumner went off to one side to hold another council. Bradley came over to where Andy Hall and I were standing and completely broke down and shed tears, and said such actions made him feel like a child again. By that time the Major came up to where we were standing and said, "Well, Billy, we have concluded to abandon the river for the present," stating that on account of the scarcity of provisions and because the rapids were getting more severe, he thought the better thing to do was to leave the river, as it could not be more than one hundred miles to some settlement in Utah, and that we would get a new supply of grub and return and complete our journey. By that time all the boys were standing and listening to him. When he finished his say, I asked him if he would sell the boat to Andy and me. He said if we would come back and finish the trip he would give us the boat. I told him I proposed to finish my part of it then.

I said, "Major, you have always looked to Hall and me as being too young to have anything to say in your council, but Hall and I are going to go down this river whether you or any of the rest go or not." And I told him that if he left the river I would not think of following him one foot on land, that my mind was set. Then the Major said, "Well, Billy, if I have one man that will stay with me I will continue my journey or be drowned in the attempt." I told him that Bradley, Hall, and I had made up our minds to continue and that I thought the worst of the rapids were passed, and that if he had taken me into his council he would have soon found out my attitude on that point.

Sumner spoke up and said, "Stay with it, Billy, and I will be with you." It did not take long to settle the rest of it. The Howland brothers and Dunn had made up their minds and would not change them. Of course, we knew what was the reason Dunn left. As for any fear, he did not possess it. And as for the other boys, they never showed any signs of fear. The elder of the Howlands had been in the boat with me since his own boat was wrecked.

We all crossed over to the north side, hid our supplies and instruments, and left one boat for the boys. It is my opinion that if the Howland boys had not agreed to leave the river in the council I referred to, they would have come with us. But they were sore about the way Dunn had been treated. I will admit that at that time they had more than an even chance to get out by land. We lifted our boats up about thirty feet over a ledge of rock and put them in a little bay in a crevice or crack between the rocks which was some twenty feet in width and thirty-five or forty feet in length. Here the water was still, but it ran like lightning in front of the crevice.

We had the two boats. The other one we left with the Howland brothers and Dunn. We again asked them to come with us. Dunn held me by the hand and tears came into his eyes as he said he hated to leave Sumner and me, that we had had many a hard and daring time together before we ever saw the Colorado River. "But," he said, "Billy, you cannot blame me." I could not answer. For once in my life I was hurt to the very heart, and in silence I shook his hand for the last time in this world. All this time Sumner and Hall were talking with the Howland boys.

The Major came into my boat, and we started first, but when we struck the main current it was so swift that it sent us back in the eddy in the little cove. By this time things were getting interesting, and again Dunn and the Howland brothers said we would never make it. I said, "Watch my smoke this time!" I told Hall to put all his strength in the oars, and I would do the rest. The Major got a firm hold with his left hand and sat down in the bottom of the boat. I headed for the lower side of the cove so as to strike the main current more on a down stream course than before. It was perhaps thirty feet from the mouth of the cove to the middle of the high waves which were over fifteen feet in height, but Hall had the boat under such headway that I could manage it with my steering oar, so I caught the side of the main waves, then cut them for the other side, which we made all right and landed below. Then came Bradley, and Sumner and Captain Powell in the other boat.

We took in perhaps only thirty gallons of water in my boat. The other boat did not fare so well, as it struck the rapid too high up, but it got through all right. We all landed and hallooed to the other boys that we left on the rock to come, but they would not.

It was here Major Powell took off his life preserver and handed it to me, saying he would have no more use for it and would make me a present of it. I told him he had better keep it on, but he said that he felt safe with any man who could come through the way I did between the rocks, and that he would make me a present of it. I thanked him, and said I would keep it to remember the Major and the daring trip and hardships down through the entire length of the Colorado River Canyons. I have the life preserver now in my possession, although it is unfit for use by reason of old age.

W. R. Hawkins

There were others besides myself who were inclined to doubt the truth and sincerity of Major Powell's rather sentimental account of the departure of the Howlands and Dunn at Separation Rapid. Among them was William Wallace Bass, of Grand Canyon, Arizona, Grand Canyon guide and profound student of the canyon's history, ancient and modern.

Seeking out William Hawkins, then the last survivor of the expedition of 1869, living in Graham County, Arizona, where he was a prominent rancher and Justice of the Peace, Mr. Bass obtained from him, just a few months before his death in 1919, a second first-hand account in his own hand writing of the Powell expedition of 1869.

The following year Mr. Bass brought out Hawkins' paper in a little booklet entitled, *Adventures in the Canyons of the Colorado,* published by the author at Grand Canyon. Written twelve years after the narrative given to me, this later version substantiates the assertions made in 1907. Because of a wealth of small details not given in the previous account, the Hawkins statement of 1919 deserves to be set forth here, with Mr. Bass' permission. Taken together, the Hawkins narratives leave little doubt that Major Powell avoided a great deal of the truth in his own written explanation of the whole unfortunate business. Beginning at his first mention of trouble brewing between the Major and Bill Dunn, here is the remainder of Hawkins' story as given to Bass:

Now our trouble begins, and plenty of bad rapids in the river. Dunn was the one who took the altitudes with the barometer, and it was here we had the first real trouble in the party, although Powell had named Dunn the "Dirty Devil." But the rest of the boys looked over that. At noon, while we were making a portage and letting the boats over a bad place, the ropes happened to catch Bill Dunn under the arms and came near drowning him, but he managed to catch the ropes and come out. While we were eating our dinner, Sumner said that Dunn came near being drowned and the Major's brother made the remark that it would have been but little loss. The Major spoke up and said that Dunn would have to pay thirty dollars for a watch belonging to him that had been soaked with water and ruined, and that if he did not he would have to leave the party.

Andy Hall and I were down at our boat, I having gone down after a cup and Andy had remained at the boat fitting one of his oars. When we returned to where they were eating, Sumner asked me what I thought of the Major's proposition, and I asked him what it was, and he then related what had been said. I asked the Major if that was his desire, and he said that it was. I made the remark that a part of his wishes could not be granted, as it was impossible to get out of the Canyon on account of the abrupt walls. He then said that it made no difference whether Dunn got out or not. I then said that I was sorry that Dunn had been jerked into the water and got the watch wet, and that I was sorry he felt that way with one of his party. The Major seemed to be offended at my remarks and said I had no right to pass on the matter, also that neither Hall nor myself, in the future of the party, would be expected to say anything, as we were too young. Hall made the remark that we had old heads on our shoulders anyway. Before this time everything seemed to be getting along fine, as each man had certain things to do, and I was doing the cooking, and I generally found plenty to do.

Our meal was ready and we all seated ourselves

on the rocks to eat our dinner. Up to this time I had always helped the Major all I could and washed his hand (as he only had one) and generally found him a good place to sit at meals, sometimes a few feet from the rest. Before this it never made any difference to me, but now it did, for, as Andy Hall would say, he raised hell with himself in the break he had made with Dunn. I could see that there was a different feeling in the whole party at this time. The Major had sat down several feet from the rest of the party. I poured out each man a cup of coffee and one for him also and we all began to eat. He then asked me why I did not bring him his dinner as I had been doing before. I told him he had just said that he was going to make a change in the outfit, and I told him that I had made that change to start the ball rolling, and that he would have to come and get his grub like the rest of the boys.

His brother then handed his dinner to him. After dinner Sumner asked him if he had changed his mind in regard to Dunn and the watch and he said he had not, and that Dunn would either pay for the watch or leave the party. Dunn, Hall, Bradley, and myself were near the boat and about twenty feet from the Major and Sumner. We could not hear what they were talking about, but we had decided that if Dunn left the party we would go with him. Of course we expected opposition to what we intended to do, so after we had talked the matter over we wanted Bradley to go and tell the Major what we intended to do. But Bradley decided I had better go and tell him myself, as I had made the plan of going with Dunn. I went to where Sumner and the Major were talking with the two Howland boys.

I told the Major that Bradley, Hall, and I had decided to go with Dunn, and that we would take my boat (the cook boat) and some grub, and would pull out, and he could come when he got ready. He said he would not stand any such work, that it would be the ruin of his party. I told him that it was all his own fault, and that I had no more talk to make, and went back to the boat. I found Dunn, Bradley, and Hall waiting to see what had happened, but before I had time to tell them, Sumner came and began to talk to us, telling us to not feel put out, that the Major was hasty, and to give him another chance. Dunn remarked that the Major did not like him anyway. If he had, he would never have named the Escalante River the "Dirty Devil."

We camped at that place for the night and in the morning the Major said he would take thirty dollars for the watch and that he could pay for it when we got through. None of the party except the Major liked Capt. Powell. He had a bull-dozing way that was not then practiced in the West. He threatened to slap me several times for trying to sing as he did, but he never did slap anyone in the party. We all moved off down the river O. K., but our provisions began to run short, and rapids became more frequent, some of them very bad. But for a few days everything went all right. The boys would tell Indian adventures at night that someone had had. The remark was made that Dunn had nothing to say, and Captain Powell said he guessed Dunn did not know much about Indians. The Major chipped in and said, "Nor anything else." Sumner took it up for Dunn because he knew there would soon be trouble. He told Powell that Dunn had been wounded four times by the Comanches, so it all passed off.

The next day we had some very bad rapids, so bad that it was necessary to let the boats around some large rocks. In order to do this, and as Dunn was a fine swimmer, the Major asked him to swim out to a rock so the boat would swing in below. He made the rock all O. K. and was ready to catch the rope which was supposed to be thrown to him, so he could swing the boat in below, but the Major saw his chance to drown Dunn, as he thought, and he held the rope. That was the first time that he had interfered in the letting the boats around bad places, and the rope caught Dunn around the legs and pulled him into the current and came near losing the boat.

But Dunn held on to the rope and finally stopped in water up to his hips. We were all in the water but the Major and the Captain. Dunn told the Major that if he had not been a good swimmer he and the boat both would have been lost. The Major said as to Dunn that there would have been but little loss. One word brought on another, and the Major called Dunn a bad name and Dunn said that if the Major was not a cripple he would not be called such names.

Then Captain Powell said he was not crippled, and started for Dunn with an oath, and the remark he would finish Dunn. He had to pass right by me and I knew that he would soon drown Dunn, as he, so much larger could easily do. He was swearing and his eyes looked like fire. Just as he passed I caught him by the hair of his head and pulled him over back into the water. Howland saw us scuffling and he was afraid Cap would get hold of my legs. But Dunn got to me first and said, "For God's Sake, Bill, you will drown him!" By that time Howland was there and Cap had been in the water long enough and Dunn and Howland dragged him out on the sand bar in the rocks. After I got my hold in Cap's hair I was afraid to let go, for he was a very strong man. He was up in a short time, and mad! I guess he *was* mad! He cursed me to everything, even to being a "Missouri puke." I wasn't afraid of him when I got on dry ground. I could out-knock him after he was picked up twice.

He made for his gun and swore he would kill me and Dunn. But this talk did not excite me. As he was taking his gun from the deck of the boat, Andy Hall gave him a punch behind the ear and told him to put it back or off would go his head. Cap looked around and saw who had the gun, and he sure dropped his. This all happened before the Major got around to where we were. He soon took in the situation and came to me and made the remark that he would have never thought I would go back on him. I told him that he had gone back on himself, and that he had better help Cap get the sand out of his eyes, and that if he monkeyed with me any more I would keep him down next time.

Sumner and I had all we could do to keep down mutiny. There was bad feeling from that time on for a few days. We began not to recognize any authority from the Major. We began to run races with the boats, as the loads were almost all gone. It was fun for the first two days, but then the water began to get rough. Hall, Howland, and myself were in my boat. I had become an expert in bad rapids. We ran several that the other two boats were let over with ropes. We stopped at noon one day to wait for the other boats. We were at the head of four bad rapids. It was some two hours before the other boats came and I had coffee all ready, as that was our principal food then. We had but little flour, but had plenty of dried applies and coffee. We lay in camp that afternoon and the Major and Sumner spent the afternoon in trying to find a place where we could let the boats over the first rapid with ropes. But they failed to find any place where we could get footing enough and the walls were too high for our ropes, so the

(Photograph by Julius F. Stone, 1909)

ROUGH WATER AT GRANITE LEDGE RAPID

Major said we would try to find a place on the west side the next day.

That evening late the Major and Sumner and the two Howland boys held a consultation (as I afterward found out) to see about leaving the river with all hands. He said we would cross over and leave our boats with instruments under some large rocks, and that we then would go out to some Mormon settlements and get some grub and return to our boats and continue on down the river. The Major asked me to bake up all the flour we had and make the bread into biscuits or dough-gods, as we called them, as flour and water was what we had to make them with. In about three hours I had them all baked. I told the Major that the bread was ready and he called the boys and told them all his intentions as to leaving the river. That was the first time Hall and I knew anything about what was going on. I told Hall to take our shares and put them in the boat. The Major said that each man should keep his own part, since we might get separated. I told him that if we had enough coming to us to pay for the boat he could keep it.

Dunn, O. G. Howland, and Seneca Howland had made up their minds to go. Dunn said he hated to leave Hall and me, as we had been together a long time, and that we would perish in the river, and that we had better come and stay with the party. I told him that was what I was doing, that I called Hall,

Bradley, and myself a party of three, and each one of us a party of one. While we were talking the Major came up to me and laid his left arm across my neck, tears running down his cheeks. By that time the rest of the boys were present and the Major said to me, "Bill, do you really mean what you say?" I told him that I did, and he said that if he had one man that would stay with him that he would not abandon the river. I simply said that he did not know his party, and that Andy Hall and myself were too young to have any say in council. I said, "We are off now." He said that it was noon, and if I would make some coffee that we would have a cup of coffee together. I have been present at many solemn occasions, but I never witnessed one that came up to this. Some strong men shed tears. Bradley said it made him a child again. We crossed over to the west side of the river and there we left our instruments and one boat.

This is the last time we ever saw Dunn and the two Howland brothers alive. Some years afterwards I, with a party of some others, buried their bones in the Shewits Mountains, below Kanab Wash. As to Powell's leaving the party at Lee's Ferry, there was no ferry on the river. No one except some Indians ever crossed. There was no place known as Kanab Wash when we first came down. Powell never left the party until we got through to the mouth of the Virgin River, where he and his brother were taken to the railroad or stage by some Mormons who lived on the Muddy. Sumner, Bradley, Hall, and myself continued on down the river. Hall and I stopped at Ehrenburg, and Sumner and Bradley went on to Yuma. From there Sumner went to Denver and Bradley to San Diego, where he died. Sumner died at Vernal, Utah, so I heard, and Hall was killed near Globe, Ariz. Powell and his brother both died somewhere in the East, and I am here nine miles below Phoenix.

CHAPTER III

JACK SUMNER'S ACCOUNT

THE following story by Jack Sumner was given to me by him about six months before his death, which occurred July 5th, 1907. I knew Sumner for some eighteen years. For a time he was associated with me in work upon the Colorado. Hence it will be proper to give a short account of the history and character of the man to whom is due so much of the credit for the success of the first real exploration of the Colorado River.

Sumner was born in Illinois in 1840. When a mere lad he moved with his parents to Iowa, then a frontier state. There on the Cedar River he became an expert boatman and hunter, and acquired a skill and daring destined to serve him well in years to come. When the Civil War began he was among the first to enlist for the short term. When this term of service expired, he re-enlisted in Company E, 32nd Iowa Volunteers, being mustered in October 6th, 1862, and mustered out May 2nd, 1865. An expert with a rifle, Sumner rendered brilliant service as a sharpshooter and scout. If in his immediate command there was a call for any extra hazardous duty, Jack Sumner was the first man to volunteer. Whatever there was of danger and privation in the life of a private soldier in the Army of the West he cheerfully shared.

In June, 1866, Sumner went to Colorado and was one of the party that accompanied Bayard Taylor in his journey through the Rocky Mountains, acting as one of the guides under his brother-in-law, William N. Byers, of Denver. In the fall of that year he located at Hot Sulphur Springs, in Middle Park, spending his time trapping, hunting, and trading with the Indians. His treatment of the Indians was kindly yet without fear. At one time while crossing the range for supplies, he was fired upon and promptly returned the shots. Next day two Indians were found dead by their companions. Soon Sumner's cabin was filled with angry savages bent on vengeance. He explained the facts to the Indians, but they were not appeased. With a coolness that excited their wonder, for there was not a white man within fifty miles of him, Sumner drew his revolver. Pointing it at a couple of boxes of powder that he had on hand, he told them the sooner they departed the better, as he would blow cabin and all to pieces. The Indians were impressed by his nerve, and at once left. After this they treated him with greater consideration. "And would you have fired into that powder?" was once asked Sumner. "Of course I would," he replied. "It would have been better than death at the hands of the Indians, and then we would have all gone to hell together."

It was at this same cabin that Major Powell found

Sumner in the summer of 1867, when he came into Middle Park with letters of introduction from Byers. Sumner was a true frontiersman, quiet and generous, and yet with a temper and spirit that knew no bounds when he was treated unjustly by others. Like most strong men, he had his faults, but they were not of the heart.

After braving the dangers of all the canyons of the Colorado, and more than once saving the life of the commander of his expedition, Sumner reached Fort Mojave, and there he writes, "After two years of hard work of exploration of the Colorado and its tributaries, I find myself penniless and disgusted with the whole thing, sitting here under a mesquite bush, in the sand, writing this journal."

Knowing Sumner as intimately and well as I did, I am satisfied that every word of his story here given is true as he saw it. In matters where there could have been an honest difference of opinion, I am again inclined to take Sumner's version, as he had no motive behind his story except to tell the truth. To any one acquainted with the frontiersman of forty or fifty years ago this story of the chief boatman of Major Powell's famous expedition of 1869 will have a peculiar interest. There were frontiersmen and frontiersmen. I have known all grades, and lived with them in their camps year in and year out. Sumner, even with his natural faults, was one of the highest type. His running comments upon men and events all over

JACK SUMNER'S ACCOUNT

the world from the time of Moses to the present day are as natural and as pleasing as the memory of an evening's chat around a campfire in the Rocky Mountains.

As I have been requested by many different people to write my experiences of sixty years on the frontier, and especially during my connection with the first Powell expedition down the Colorado River in 1869, I offer this crude, ill-written tale to the gentle reader. If he does not like it, he can dispose of it as he sees fit. Every word of it is true, but as my journal has been lost for many years * I cannot give exact dates of the Colorado River exploration trip. Yet the incidents here described are as fresh in my memory as those of but a few days ago. I will try to describe the Powell exploring expedition from its first inception to its close.

In the fall of 1867 Major J. W. Powell came to me at my trading post in the Hot Springs in Middle

---

* The history of Sumner's original journal kept in 1869 is as follows: On reaching Fort Mojave, Arizona, Sumner says, he made a complete copy of his journal at Major Powell's request, and sent it to him in Washington. This was in existence for some years and was seen by Dellenbaugh in 1871. This first copy as described by Sumner cannot now be found in the archives at Washington where other records of the expedition are kept. A mutilated copy supposed to be a copy of this first copy was found by me in the spring of 1907 among Major Powell's papers and on Sumner's order was turned over to me, and is now in my possession. This second copy is not complete, and is not dated. It is not known who made the copy, but it is known to have been in existence for quite a number of years. Sumner's original notes, kept each day during the expedition, were lost years ago.

Park, Colorado, bringing letters from Denver parties requesting me to show him the country at large, and give him all the information I could, especially in regard to the natural history of the Park and adjacent country. I saddled up and took him around to about all the points of interest, making daily collections of the various animals at that time inhabiting that part of the Rocky Mountains. They were elk, mule deer, mountain sheep, antelope, and the various fur-bearing animals, three kinds of bear—the grizzly, the cinnamon, and the black—beaver, otter, marten, and mink; the grey wolf and his bastard cousin, the coyote, wolverine, silver cross and red foxes, besides some smaller animals generally classed as vermin by a free trapper, of no value for furs, but of great interest to the naturalist.

In our evening talks around the campfire, I gave the Major some new ideas in regard to the habits of animals, as he had gotten his information from books, and I from personal observation of the animals themselves, in a perfectly wild state. We differed widely, especially so in regard to beaver and how they worked. I had to take him to a beaver village on the sly and let him see for himself before I could get that silly idea out of his head that beaver use their tails as trowels and sleds in building their dams. He saw at once that they used the tail as a sculling oar, and for nothing else, and bring the mud for their dams in their mouths when they have to go any depth under water for it. In shallow water they roll up a batch of mud and take it to the place wanted by rolling it with their front feet —if a large piece by moving backward, if a small piece, by carrying it between their paws and chin.

Another pet notion of his was that grizzly bears must surely be grey in color. As a matter of fact, they range from jet black all over to as light grey as the grey wolf. The only marked difference between the grizzly bear and other bears is in the claws. The black and cinnamon bears have retractile claws, like a cat, and are excellent climbers. The grizzly cannot pull his claws up out of the way and cannot climb a tree. If a novice is trailing a bear and sees claw marks six inches long in front of the foot, I would recommend that he use extreme caution, as a grizzly has a bad habit of doubling back and concealing himself near the trail, ready to pounce on anything following him up. If a hunter has a deadly twenty-two, I would advise him to go home at once.

After spending two or three weeks wandering in the Park, the Major seemed to get stuck on me for some reason or no reason at all, and wanted me to leave Middle Park, and go with him the following summer to the Bad Lands of Dakota on a geological trip. I declined the proposition and fired back at him the counter-proposition—the exploration of the Colorado River of the West, from the junction of the Green and Grand rivers to the Gulf of California. He at first scouted the idea as foolhardy and impossible. I urged on him the importance of the work, and what a big feather it would be in our hats if we succeeded. After several windy fights around the campfire, I finally outwinded him, and it was agreed that he should come out the following spring and we would make the attempt. I believe Major Powell states in his report that the exploration of the Colorado River had been in his mind for years. He mentioned nothing of the kind to me previous to our discussion and agreement. The idea was certainly not his own.

Collecting the specimens I had gratuitously gathered for him in the Middle Park region, Major Powell departed for the States, agreeing to be back as soon as the grass started the following spring, and we would then commence the generally supposed foolish and impossible task. Each of us was to bear, share and share alike, all of the expense. I remained in Middle Park that winter collecting for the Smithsonian Institution.

Major Powell appeared on the field at Berthoud Pass the following June (1868), with a gang of twenty-five or thirty college students from fifteen to twenty-five years of age, with no experience whatever. They were good enough in their place, but about as fit for the work supposed to be ahead of us as I would be behind a dry-goods counter, or hell for a powder house. He then told me he had changed the original plan of attack, as he had secured authority from Congress, through the influence of Senator Trumbull of

Illinois, to draw supplies for twelve men at any western army post whenever the supplies were called for, which was a pretty good thing, as supplies were very expensive at all frontier posts.

After remaining at Berthoud Pass for two or three weeks, the outfit moved down to Hot Sulphur Springs and made that place headquarters for the remainder of the summer, collecting, mapping the country, and measuring various mountains around the extreme head of Grand River, and made the first ascent of Long's Peak, August 4th, 1868. In October we commenced moving everything to the White River country. We had considerable trouble getting all our plunder over. We had to pack in relays, as there was a lot of stuff of no use whatever. As game was plentiful on the route, we lived well and finally got bag and baggage to the White River and built winter quarters on the north side of the White River at the point afterwards made historical by Nathan Meeker, who built an agency there for the Ute Indians.

Later on Meeker and all the men of the agency were killed and all the women kept prisoners for some time, and they would have been killed, too, but for the influence of Chief Ouray of the Uncompahgre band of Utes. A similar fate was narrowly averted for our party of '69 through the friendship of Chief Antero of the Uinta tribe of Utes and myself. Major Powell had measured off some geological work and set some stakes. The Utes thought that meant farming, and of course they would not have it that way. As we were only ten to about a thousand, our cake would have been dough if they had not been well talked to and the obnoxious stakes pulled up and thrown into the White River.

It has always been a strange thing to me that educated men seldom use common sense in their dealings with wild Indians. Meeker proposed to make the Utes good farmers and highly civilized in a few days. We know the results. Powell had a notion that his name and prestige would carry him anywhere, a common and very often a fatal mistake with army officers. Note the cases of Col. Fetterman and Gen. Custer. I doubt very much if Lewis and Clark could have traveled across the continent but for the cool head and quiet ways of Clark in dealing with the Indians. Fremont would have been stopped at the forks of the Platte but for the cool heads of his guides and scouts. In all his wanderings you can see the influence of Godey, Carson, Bridger, and a dozen other frontier men. Read Fremont's report and see if I am not right.

We camped on White River from November, 1868, until next March, collecting and exploring the country for a hundred miles around, the snow being too deep for us to make any very extended trips. Major Powell's students had by this time dwindled down to one out of the twenty-five, and he was terribly homesick. March 4th, 1869, we abandoned our winter camp and started through pretty deep snow for Bear River, about sixty miles northwest on the old trail to Brown's Hole. We reached Bear River in three days and found no snow at all, good grass, and plenty of antelope but no other game.

Next day the main party pulled on, Major Powell going back to the States to make preparations for our Colorado River journey. Howland, Dunn, and I remained behind to examine the country adjacent to Bear, Little Snake, and Vermillion rivers. We spent several days in the work and then moved on to Brown's Hole, where we camped for about two weeks and had lots of fun. Deer were very plentiful and the water fowl were in flight, so we had duck soup and roasted ribs about every meal. After we got tired of our camp in Brown's Hole, we proceeded on our way to old Fort Bridger by way of Henry's Fork, seeing nothing of interest but a fight between a large flock of mountain ravens and a mountain lion in the cliffs of a cavernously eroded butte—a form of butte characteristic of that graveyard of prehistoric reptiles. We reached Fort Badger, and then on to Green River Station on the Union Pacific railroad, where we camped and awaited orders and in the meantime tried to drink all the whiskey there was in town. The result was a failure, as Jake Fields persisted in making it faster than we could drink it.

About May 12th, '69, Major Powell returned from the States with the four boats which had been built in Chicago from plans I drew at our winter quarters on

White River. Three of them were twenty-one feet long, six feet wide and two feet, two inches deep, decked fore and aft about five feet, and with watertight compartments. The pilot boat, the "Emma Dean," was sixteen feet long, four feet wide, and twenty inches deep. Major Powell brought out a lot of necessary trinkets, and also a young scientific duck who was not at all necessary. However, he did not give us much time to imbibe his wisdom, as he stayed only one day. One good look at the Green River and the gang was enough. He vamoosed the camp that night, and we were left in darkness, mourning bitterly. I then induced Seneca Howland, Frank Goodman, and Andy Hall, the latter a rollicking young Scotch boy, to cast their lot with us. George Bradley, a sergeant from Fort Bridger, also joined us at Green River, he having been discharged by order of Gen. Grant especially to go with us. He was something of a geologist and, in my eyes far more important, he had been raised in the Maine codfishery school, and was a good boatman, and a brave man, not very strong but as tough as a badger.

We recalked and repainted the boats and, having everything in readiness, on May 24th, 1869, we pulled out into the swift current of the Green River in the following order: the "Emma Dean" with a crew of J. W. Powell, Bill Dunn and Jack Sumner; "Kitty Clyde's Sister," Walter Powell and George Bradley; "No Name," O. G. Howland, and Frank Goodman;

"Maid of the Canyon," Bill Hawkins and Andy Hall.

The boats were ordered to keep one hundred yards apart. Flag signals were arranged as follows, always to be given by Major Powell from the pilot boat: flag waved right and left, then down, "Land at once"; waved to right, "Keep to right"; and waved to left, "Keep to left of pilot boat." As Major Powell was the only free-handed man in the outfit, he was supposed always to attend to that part of the business, but I fear he got too badly rattled to attend to it properly on several occasions, notably so at Disaster Falls in Lodore Canyon. As Bradley and I were the only experienced boatmen, there were some ludicrous mishaps for the first few days, running aground at the head of sand bars being the principal trouble. Hall, who had once been a bull-whacker, swore that his boat would "neither gee nor haw nor whoa worth a damn" —in fact, it "wasn't *broke* at all!"

The first day we ran down about ten miles and camped on an island. It rained most all day and night. We pulled out next morning and rowed nearly all day and arrived at the mouth of Henry's Fork, a famous beaver stream coming in from the northwest, and camped at the head of Flaming Gorge, our first canyon to butt into. The walls are not very high and are of a brilliant, fiery red color. We had no trouble in running the various canyons above Brown's Hole, but had to make a portage at Ashley Falls which kept us busy for nearly a day, as all the large boats were loaded to the safety limit. We carried the supplies over the rocks on the north side about fifty feet above the river. The boats were then let down by lines and rounded in and landed at the foot of the falls. It was hard work, but there was no danger in it to either men or boats. In such expeditions it is always better to be safe than sorry.

Passing the falls, we leisurely rowed down and camped in about the middle of Brown's Hole, a fine camp, where we enjoyed ourselves to the limit, fishing, shooting ducks, and geologizing. From the camp we dropped down about ten miles and camped in the head of Lodore Canyon, measured the cliffs and took observations for exact location. We camped there one day and then attacked the worst place on Green River, as we soon found to our sorrow. There is a nest of nearly continuous rapids for twenty miles that is a holy terror. We had some trouble from the start in Lodore Canyon, shipping more or less water at every rapid.

About ten miles from the head of the canyon we encountered what was afterwards named Disaster Falls, a very bad rapid, or rather four very bad rapids. As I was always in the lead, I stood up on the forward deck and saw at once our danger, and immediately landed on the left bank. Two of the other boats following did the same, but the "No Name," in charge of O. G. Howland, failed to make a landing It was drawn into the rapid and ground to pieces on the

(Photograph taken in 1905)

JOHN C. ("JACK") SUMNER
Chief Boatman of the first Powell expedition

JACK SUMNER'S ACCOUNT

rocks below. The three men escaped by a scratch on a small sand bar island near the right bank. To get the men off was a conundrum for sure. The channel was not wide, but the river ran with the speed of a race horse, with a fall and a cataract below that meant certain death to anyone carried into it.

The three men were finally rescued by a man* who took the "Emma Dean" across the channel, making a landing only by a scratch. The boat was then pulled up to the extreme head of the island and the men taken aboard. Major Powell, in describing the rescue, makes a mistake in saying all hands pulled on the oars for dear life. The man who took the boat over would not allow any oars in the boat but the oars he used. When all were on board he compelled the three men to lie down flat in the bottom of the boat, and he rowed the boat back alone, as he did not care to risk certain destruction to all by a false stroke. The trip was made all right and the men rescued, none the worse for it except for a good ducking. But a good boat was gone and a full load of supplies.

This wreck marked the beginning of the many quarrels between Major Powell and O. G. Howland and Bill Dunn. As soon as Howland got out of the boat after the rescue Major Powell angrily demanded of him why he did not land. Howland told him he saw no signals to do anything, and could not

* The heroism here so modestly described by Sumner and credited to "a man" was that of Jack Sumner himself, as testified to by Major Powell, page 25 of Report, 1875.—R. B. S.

see the other boats that had landed until he was drawn into the rapid, when it was too late. I asked Hawkins and Bradley in charge of the other boats if they saw signals to land, and they said no signals were given, but as they saw me turn in they suspected something wrong and followed suit at once.

We finally worked our way through Lodore Canyon and reached the mouth of Bear River (Yampah), the largest stream joining the Green in its whole length. We camped there two days, repaired everything, and took observations. We caught fish wholesale, some of them pretty big fellows. They are called Colorado salmon, but do not resemble a salmon any more than I resemble a dude, and are about as edible as a paper of pins cooked in lard oil.

I believe Major Powell speaks in his report of finding a bake-oven and a lot of tin plates at the foot of Disaster Falls, and concludes they were the relics of Ashley's party. They could not have been, for several reasons. I have been among different trapping parties, and I never saw a bake-oven in any of them, they being too cumbersome to carry. Old "Cut Rocks" Brown, as he was called, trapped that section of the country pretty thoroughly many years ago, but he was never known to have any cooking utensils but a battered brass kettle holding probably a gallon, so they could not have been his. Brown was a queer old cuss, always wanting to get off alone in some canyon, "cut rocks," as he called them, to conceal himself from the red-bellies. Brown's Hole is named for him, but he certainly left nothing there except, possibly, some half-breed Ute children. The camp utensils were probably left by some of the bullwhackers who made Brown's Hole a wintering place for the many work cattle used in transporting supplies for Johnston's Army during the Utah war and between the time of the Mountain Meadow Massacre and the beginning of the Civil War. When they went into winter quarters, some of the more adventuresome of these men hunted and trapped to make a few dollars on the side, as their wages were cut pretty heavily during winter quarters. Some of them went down there to trap otter and were too lazy to pack back the camp outfit. The regular trappers seldom wintered in places where they could not take their horses.

Jim Baker told me that he and Sam Anderson went in bull-boats down Bear River Canyon to the Green and down the Green River to Kelly's Hole. Here they weakened and walked back to the forks of Little Snake and Bear Rivers, where they had left their horses. Baker was a thoroughbred frontiersman of long standing who ran away from his home near Springfield, Illinois, went to St. Louis and joined the American Fur Company in 1831, and started for the Rocky Mountains at once. He was as gentle as a child when used rightly, a wounded grizzly bear when provoked by some petty act of meanness in any one, a handy man with his fist, and one who would not hesitate a second to use knife or rifle if they were handy. He died about five years ago [1897 or '98] and is buried on a bluff near the forks of Little Snake and Sanery Rivers. The words "Friend and companion of Fremont" are cut on his tombstone. Baker always advised me not to attempt the Colorado River, as he considered it an impossibility and was sure there were heavy falls, and that if any of the warlike Indians saw us they would be sure to cut us off while making a portage. I have often thought what a nice kettle of fish we would have been in had the Navajos or Havasupais attacked us from the cliffs while we were making a portage at some bad rapid.

Breaking camp at the mouth of Bear River and pulling round the beautiful cliffs of Echo Rocks, we worked our way down to a fine circular park called by the trappers Kelly's Hole, after a hunter of the American Fur Company. I believe it is now called Island Park. I don't remember any portages on the way, but all hands were ducked. Kelly's Hole was an ideal winter camp, swarming with beaver and, in the winter, thousands of mule deer and some mountain sheep. We camped in the middle of the park and had fine sport shooting half-grown wild geese with our pistols. Hawkins killed a fine mule deer. We were amused at Hall trying to kill a loon. After firing a dozen shots at it, he blurted out, "Damn you! If I could not kill you, I filled you so full of lead you had to sink and drown!"

It is generally supposed that this wary bird, the loon, can dodge a bullet. That is not the case. They sit so low in the water that the hunter overshoots them, nine times out of ten.

I presume Kelly's Hole is now ruined by the sheep-herders, who have overrun every place they can get their sheep into, which soon destroy every green thing, root and branch. It seems there is no place left in its primitive beauty. Go where you will, you will find the sheep-herder or the cattle baron who claims the valley and all the surrounding country, and who always objects to a prospector's or trapper's picketing his horse.

Breaking camp in Kelly's Hole, we worked our way down through a low canyon and entered the large and beautiful Uinta Valley and had a fine time for the next three days floating on the wide, gentle current of Green River, shooting many water fowl of all kinds, and shooting *at* antelope and that nuisance of the world, the coyote, which frequently came trooping down to the bank to view the strange things going down the river.

About four o'clock the third day we reached the ford on Green River, a short distance above the mouths of the Uinta and White rivers. The Uinta comes in from the Uinta and Wasatch Mountains to the northwest, and the White comes in from the plateau region to the east in Colorado, called the Trapper's Lake Country. Near our camp there seems to have been a favorite crossing for the Indians for centuries, and for many years for the whites. There is a good ford in low water. The river is so wide there is but little current in high water, and consequently it is easy to swim. Captain R. B. Marcy crossed here when he was sent from Salt Lake to Fort Massachusetts (now Fort Garland) by order of General Albert Sidney Johnston, later killed at Shiloh when in command of the Confederate army. It was in this same battle that Major Powell lost his right arm. Bill Hickman, the noted Mormon Danite chief, captured one of Johnston's supply trains, burned the wagons and supplies, and ran off the cattle, thus reducing the Utah army of invasion to such straits that General Johnston was compelled to send Captain Marcy across country in the winter to the head of the Rio Grande to carry dispatches and get supplies, Fort Leavenworth being too far off. There were about one hundred men in the company, under the guidance of Jim Baker. They had a pitiful time of it indeed, living on the mules as they died in the snow. Jim Baker often told me that he couldn't look a mule in the face after that trip.

Major Powell and his brother, Hall, and Frank Goodman left camp and hoofed it up to old Fort Roubidoux, or as it was then called, Fort Winter, a trading post at Disaster Falls. The Howland brothers, Dunn, Bradley, Hawkins, and I remained in camp and amused ourselves by taking astronomical observations, shooting ducks, and fishing. Major Powell was gone five days, and brought back a shirt-tail full of supplies. I thought at the time it was a damned stingy, foolish scheme, as there was plenty of supplies to be had, to bring back such a meagre mess for nine to make a thousand-mile voyage through an unknown canyon, but as I wasn't boss I supposed I ought to keep still about it. Goodman, having had all the experience his health called for, stopped at the post. He had had several close calls, and possibly ran out of nerve. He was a fine singer of sea songs, and we missed him around the evening camp.

The next morning, July 6th, after the arrival of the rest of the party, we pulled out and went down the river about ten miles to Johnston's Island, where a squaw-man had planted a garden. Having landed, Hawkins and Hall started to investigate, and as they could not find anything else, they proposed to have a mess of greens. They proceeded to filch turnip tops, beet tops, potato tops, and God knows what else. They brought a backload of the beastly stuff, dumped it into the boat, and we ran down the river a few miles and landed again for dinner. Hawkins cooked a kettle of his plunder, poured it into a gold pan and yelled, "Grub pile!"

We proceeded to devour it to show our appreciation of his ability as a first-class cook. We had not gone a mile after dinner until all hands, the cook included, wished Johnston and his garden in the middle kettle of a lower world. Such a gang of sick men I never saw before or since. Whew! It seems I can feel it yet. I remarked to Hall that I didn't think potato tops made a good greens for the sixth day of July. He ripped out an oath or two, and swore he had coughed up a potato vine a foot long, with a potato on it as big as a goose egg. Hall was somewhat given to exaggeration, and he might have stretched the matter a bit. I don't know, for I was too busy bossing my own unpleasantness to attend to his. We ran down about ten miles and stopped to recover from our mess of pottage. Having located in rather a low place, we had a busy time until midnight fighting mosquitoes.

Next morning we soon got into a nest of rapids in Desolation Canyon. There was none so very bad, but bad enough to duck us every hour. In the worst rapid the "Emma Dean" was swamped [July 11th] and instantly capsized, two fine rifles being lost, along with some bedding. We broke many oars and most of the Ten Commandments. Major Powell said he lost three hundred dollars in bills. I lost my temper and at least a year's growth—didn't have anything else to lose. Through the valley from Desolation Canyon to the mouth of San Rafael River it is very much like the Uinta Valley, the river bottoms being pretty wide and covered with a dense growth of cottonwoods and willows. The whole river for thirty miles seemed to be swarming with beaver. We shot several with our pistols as they bobbed up near our boats. I had the good fortune to get two otter out of a bunch of five as they swam past puffing and spitting like a whole nest of tom cats. Part of the beaver we cooked and ate, but as they had been living on willows they were not very palatable. There were plenty of wild geese, but so very wild that we got but few to eke out our scanty rations.

Another chapter of the Powell-Howland squabble was commenced as we left the camp near the San Rafael—a sad, bitter business. I wish I had put a stop to it long before I did. Things might have ended differently.

Winding our way carefully through very crooked canyons [Labyrinth and Stillwater], we arrived at the junction of the Grand and the Green and the head of the terrible Colorado of the West. The two streams are nearly equal in flow. The Grand is much wider and smoother at its mouth than the Green. The Green has a peculiar, vicious appearance, narrower than the Grand, and with an ugly yellowish color and a swift twisting, angry current. It seems to impart its looks and nature to the Colorado all the way to the Gulf of California.

We camped there at the junction three days—July 18th, 19th, and 20th—and overhauled our boats, supplies, and instruments. I refilled the barometer tubes and boiled the mercury in a little wikiup which I constructed of cedar boughs, using an open fire of dried willow twigs. I cleaned the chronometer, trued up the sextant, and made all snug in that line. Then I repaired the boats. But when I came to the commissary I was up a stump. We had about five hundred pounds of flour and a little bacon and dried apples. After examining the flour I found it a miserable mess of green fermentation. There was nothing left to do with it but to sift it through a mosquito netting onto a wagon sheet and let it dry. After Howland and I had sifted it we had but two sacks, presumably two hundred pounds, a not very liberal supply, considering our position and numbers at that date. But as there was no help for it, we looked the subject square in the face. There was only a dim hope of replenishing the commissary by finding some deer or mountain sheep on the way.

Off again early on July 21st, and here began the real work of exploring the Colorado itself. The first five or six miles was smooth channel. Then we came to our first bad rapid, finding an even worse one 200 yards below. The next few days we made some heavy portages over the worst rapids we had seen so far. In one place we made four portages in three-quarters of a mile. We saw driftwood thirty feet upon the rocks. God help the poor wretch who is caught in this canyon [Cataract] during high water!

As my journal has been lost for thirty years, I cannot give exact points, rapids, or incidents through

Cataract Canyon. At noon on the eighth day on the Colorado (July 28th) we rowed into camp just below a side stream coming in from the north which stinks bad enough to be the sewer from Sodom and Gomorrah, or even hell. I thought I had smelt some pretty bad odors on the battle field two days after action, but they were not up to the standard of that miserable little stream which I dubbed the "Dirty Devil" more than a generation ago. I am sure I beg the Devil's pardon, and I won't do it again if he will overlook it this time. And yet the source of that stream is in as pretty a mountain lake as ever the sun kissed, and swarming with trout.

While Major Powell and I were taking observations, some of the boys panned out a pan of gravel, and got a number of colors of gold, the first we had found. From the mouth of the North Wash just below the Dirty Devil all of the gravel bars contain placer gold, at least as far down as Paria Creek [Lee's Ferry]. It runs all the way from a few cents to as high as several dollars per yard in places. The gold is all of high grade, but very fine and therefore most difficult to save.

Through Glen Canyon from the Dirty Devil to the San Juan River there is nothing dangerous at a middle stage of water, but there is at very low or high water. In high water there are a lot of very dangerous whirlpools that will catch a boat and whirl it round till it makes the occupants dizzy and, if the boat is drawn under the vortex of the whirlpool there is not one chance in a hundred of its ever rising again. I remember one little incident or experience I had in one. The "Emma Dean" was caught in so severe a whirlpool that Dunn and I could not pull out of it to save our lives. It spun us round like a roulette wheel. I thought I saw a chance to get out as the boat spun round past a rock about thirty feet away. Seizing the rope in my teeth, I made a desperate plunge. Good lively swimming, combined with the momentum of my jump, enabled me to make the rock by a scratch. Seizing it with a death grip, I was able to pull the bow of the boat out of the swirl, whereupon it shot ahead like a scared rabbit.

In low water there is a stretch of canyon from what is now known as Hall's Crossing that is very bad for thirty miles, as the river bed is formed of a solid, uneven mass of rock, full of bumps, and where the very shallow water runs like a bat out of hell. There are four reefs that cross the entire river, and there is not more than a foot of water on them. This is now known as Shock's Rapid. I passed over it down and up in the winter of 1898, and had a soul-crushing time of it. There was with me a Mormon bishop who declared such work would debar him from the Church.

In our first trip in '69, for some unaccountable reason we did not notice the largest stream on the north side of the Colorado, below the Dirty Devil. I refer to the Escalante Creek. It breaks out through rather straight walls on the north side, and there are so many big rocks in the mouth of the creek it can hardly be defined. Besides, it makes a bad rapid immediately below. We were probably spying out the route through the rapid, and so overlooked the creek.

Arriving at the mouth of the San Juan River [July 31st], we camped there a few days. Here I had another job of repairing instruments and cursing swarms of minute gnats, so small that they could hardly be seen but as full of venom as hell fire, or as a politician is full of tricks. Just above the San Juan we saw the first prominent "Moqui" houses, though we had seen some ruins farther up the river. These people were a strange folk. Remarkable that such frail things as corn shucks and leaves should last longer than the history of a nation!

From the San Juan we moved down a few miles and camped again to correct the topographical work, as Howland had got things somewhat confused. Here occurred another spat between him and Major Powell. Will it never cease? Such silly business indulged in by educated men is liable to create a bad impression in the minds of frontier trappers. Military martinets and civilians very often disagree, however.

Dropping down from Music Hall camp, we experienced nothing worthy of note excepting daily duckings and continuous fasting. We arrived at Paria Creek, a little stream coming in from the north.

We camped there a day. Trails and ancient camps show that this point has been used for ages as a crossing place of the Colorado. There are great numbers of milling stones used by the Indians to grind, or rather pound, piñon nuts and grass seeds, mixed with dried grasshoppers and an occasional lizard, to make their daily—or rather weekly—bread. There is an ancient fort crowning a small butte near the crossing. Did the Moquis make a desperate stand here before they were wiped out by the northern hordes that swept everything before them from Puget Sound to the plains of Mexico? The invaders left barely enough of the Cliff Dwellers for seed. Their descendants, I believe, are the present Pueblo Indians, but I don't pretend to *know,* and I can't find any one to give me any better information.

We saw fresh moccasin tracks at various places in the little valley, a fact which set me on the keenest watch to avoid a surprise. Indians rarely attack an unknown enemy unless they can surprise him. We were in no condition for a fight or a foot race. If the reds saw us, as they probably did, they could see plainly enough to satisfy them that to attempt to surprise us would be very hazardous. Accordingly, none showed up.

Below Paria Creek we soon encountered some pretty bad rapids. The rapids themselves were not so bad, but there was a vicious undertow and backwater at the tail of all of them that was a holy terror. If a boat kept the dividing line between the main current and the back current it went through all right. If the bow caught in the back current the least bit, the boat was whirled around quick enough to take the kinks out of a ram's horn. The boat was sure to fill with water and was usually capsized.

In such a place in 1889 Frank M. Brown and two of his men were lost, and the rest of the party under Robert B. Stanton had to retreat by way of Kanab. Stanton reorganized and afterwards made the trip all right. I met him and his party at the mouth of Hanson Creek at the base of the Henry Mountains in December, 1889. He stopped the outfit and we had a fine game of talk of several hours' duration, he getting all the pointers I could give him, and I getting a plug of tobacco from him which I greatly appreciated, as I had been out for a month. And with me it is out of tobacco, out of humor.

Near the center of Marble Canyon is the prettiest spring I ever saw. It comes out of a straight marble wall, four or five hundred feet high, through a round hole probably six or eight feet in diameter and pitches into the river in a magnificently graceful curve of white foam. It does not seem to be permanent in its flow, since Stanton told me a few years ago that when he passed the place there was but little water coming from the hole.

From Paria Creek [or Lee's Ferry] to the Rio De Lino, Flax River or Little Colorado, there is marble enough of all kinds to build forty Babylons, walls and all, with enough to spare to build forty other cities before it would be missed. If geology is true—and it certainly is, if anything is—what vast ages the little insects must have worked to furnish the material for two thousand feet of marble! And what a length of time it took to form the miles of lime and sandstone that overlie the marble! And again what vast ages after the upheaval of the Buckskin Mountains that formed the interior lake that covered what is now southwestern Colorado and southeastern Utah, with but three small groups of porphyry mountains poking their heads above the fresh water sea!* And then, how long did it require the Colorado River to cut its channel through all the sedimentary rocks and twelve hundred feet and more into the Archean formation? Who knows? The testimony of the rocks cannot be impeached. I think Moses must be mistaken in his chronology as recorded in Biblical history. From Green River, Wyoming, to the Virgin River is one continuous geological book. You can turn leaf after leaf from the Quatenary to the Archean. Whoever can read it is a master for sure.

After about a week of daily duckings and some dangers we passed Marble Canyon and arrived at the mouth of the Little Colorado, a miserably lonely place indeed, with no signs of life but lizards, bats,

---

* NOTE: This was the Geological history as expounded by Powell and Dutton, but which has been modified since. R. B. S.

and scorpions. It seemed like the first gates of hell. One almost expected to see Cerberus poke his ugly head out of some dismal hole and growl his disapproval of all who had not Charon's pass.

We camped near the mouth of the Little Colorado one and a half days—made repairs, took observations, and checked up our records. Walter Powell took a barometer and climbed the cliff between the two rivers. We could see him as he reached the top. He looked like a mote in a sunbeam. Major Powell climbed the cliff below the Little Colorado and was gone so long I was afraid he was lost, as he went up an old Indian trail, and would never carry any arms. As we were on the edge of the Apache and Havasupai Indian country, his act was foolish, to say the least. However, he returned about dark and reported seeing nothing worse than two large rattlesnakes.

The Little Colorado at its mouth at low water is an insignificant stream. Half its volume is slime and salt. We cut loose the next morning and dropped down five or six miles on a comparatively quiet river.

It was not long before we could plainly foresee trouble, as the formation was rising and we knew that the next leaf or chapter would be granite or gneiss. We were not long held in suspense. Sure enough, as we rounded a curve—biff! up came the old Archean granite with its usual trimmings, a very bad rapid. And if any men ever did penance for their sins, we did a-plenty for the next two hundred miles. To add to our troubles, there was a nearly continuous rain and a great rise in the river that created such a current and turmoil that it tried our strength to the limit. We were weakened by hardships and ceaseless toil for twenty out of twenty-four hours of the day. Starvation stared us in the face. I felt like Job: it would be a good scheme to curse God and die, but, like him, I did not do it. Frequently something laughable would turn up to drive away the blue devils. After a day's run was made, I had to take the sextant and watch for the stars, while the Major took the time, and made the records. At times it would be well on towards morning before I could get a shot at a stray star through the small slit overhead.

About fifteen miles below the Little Colorado the first bad rapid occurs, in what I wanted to name Coronado Canyon. Major Powell told me it should bear my name if he got through and ever had the opportunity to place it on the Government map. Well, he got through all right, but he forgot his vows and named it Grand Canyon. The name is appropriate enough, but as there are a dozen "Grand Canyons" in the southwestern part of the United States, the name is liable to mislead the stranger.

Landing at the head of the first great rapid in the granite, we soon saw that it would not do to try to run it. So we let down with lines. The work was very hard and trying, as the water was icy cold and full of rocks. Seeing there was no help for it, all hands worked manfully and we got through it a little before night. Then we proceeded a short distance and ran into another bird of the same foul brood. We camped at the head of it and made the best of it for the night.

Major Powell spent the following day geologizing, as he was a nuisance in the work of portaging. His imperious orders were not appreciated. We had troubles enough without them. There was another spat between the Major and Howland at this point. Tackling the next rapid as soon as Hawkins, the cook, had called coffee and sour dough and we had disposed of it, into the rapid we went, and ran it all right. Two oars were broken, and the side of the "Emma Dean" was cracked. We repaired this damage, and hunted another rapid. That did not take us long, for they are plentiful and easily found in the Grand Canyon. And thus it went—a ceaseless grind of running or letting down rapids with lines, varied in places by making portages of boats and contents. The contents were a small item, but the boats, water-logged and very heavy, taxed our strength to the limit.

We finally encountered a stretch of water and canyon that made my hair curl. I don't know how it affected the other boys. The walls were close on both sides, with a fall of probably thirty feet in six hundred yards, a white foam as far down as we could see, with a line of waves in the middle, fifteen feet high.* And as the canyon curved to the left and shut out the view below, we were puzzled for awhile. I had charge of the leading boat, the "Emma Dean," from the start at Green River. Owing to the petty quarrels between Major Powell and the Howland brothers and Dunn, I had undertaken much of the running part of the expedition. I decided to run it, though there was a queer feeling in my craw, as I could see plainly enough a certain swamping for all the boats. But what was around the curve below out of our sight?

A fall below meant certain destruction to all. I stated the case and asked, "Who follows?" I can still hear the ringing voices of Hawkins and Hall: "Pull out! We'll follow you to tidewater or hell!" Carefully fastening the hatches and directing the other boats to keep a hundred yards apart, I started out. The "Emma Dean" had not made a hundred yards before an especially heavy wave struck her and drove her completely under water. Though it did not capsize her or knock any one out, the wave rendered her completely unmanageable. Dunn and I laid out all our surplus strength to keep her off the rocks, while Major Powell worked like a Trojan to bail her out a little. We managed to get her into a little cove with a flat shelf of rock as large as a table, where I landed

* This is the noted "Sockdologer" rapid. In his original Journal Sumner says the fall was 30 feet, exactly as did Major Powell in his field Journal. Writing nearly 40 years later, Sumner still remembers it as 30 feet, and does not increase it to "75 or 80 feet," as Major Powell does in his report in 1874. R. B. S.

instanter. I held the lightened boat until the Major and Dunn bailed her out. Then I jumped aboard and we were after the other two boats that we had seen a moment before pass like a whirlwind, full to the guards, but right side up. Away we went as if we had important business below, as in fact we had.

The other boats had disappeared around a curve, and had encountered we knew not what. Giving the "Emma Dean" all the speed we possibly could, we passed through the rapid, ducked at every wave, but as we struck them just right we did not fill. After half a mile of such work we caught up with the other boats, landed in a quiet cove. They greeted us with a ringing cheer as we rounded in. We compared notes, Hall in a rollicking way saying he felt sorry for us as they passed by, and regretted that he hadn't a picture of us.

Howland was very grave and silent. I have often asked myself this question: did he have a premonition of his death soon to come? I have been in a cavalry charge, charged the batteries, and stood by the guns to repel a charge. But never before did my sand run so low. In fact, it all ran out, but as I had to have some more grit, I borrowed it from the other boys and got along all right after that.

After running probably another mile we encountered another rapid, or rather a fall—the only direct fall we had encountered on the river. It was a direct straight fall, reaching entirely across the river, with a drop of about eight feet. As we could see there was no great danger below, I determined to run it, a decision that all agreed to. So out we pulled and made for it full speed, and jumped it like jumping a hurdle with a bucking horse—and didn't ship enough water to moisten a postage stamp. The fall gave off a peculiar sound at intervals of about ten seconds that sounded precisely like a minute gun at sea. Hence it was called Minute Gun Falls. And so it went—rapids, daily duckings, and "heap hungry" all the time. Through hard usage all the boats were getting short on oars. We therefore looked anxiously for a drift log to make them from. The river is so terrific it seems to smash everything into pieces, leaving nothing large enough to make an oar.

By great caution and good luck we reached Bright Angel Creek, a small clear stream coming out of the Kanab Plateau region, where we turned in and camped one and a half days. We found here some red spruce logs out of which oars were made by ripping them out with a common hand saw, considerable of a task. The creek is about ten feet wide in low water and a foot deep, clear enough but in August warm as dishwater. We saw no indication that white men had ever visited the valley of the Bright Angel before. There were plenty of old Indian camps, milling stones, and wikiups. Here was lost the last remnant of our soda. Hawkins put it on the bank in a gold pan. The bank caved in and away it went, pan and all. After that we had rotten flour mixed with Colorado River water, not a very palatable mixture. But after eating the "sinkers" we *knew* we had dined!

After making oars and inspecting everything, we pulled out again for more of the Great Unknown. There was not much talk indulged in by the grim squad of half-starved men with faces wearing that peculiar stern look always noticed on the faces of men forming for a charge in battle. It is not handsome by any means, but all the same it has a fascination about it that always attracts attention.

This part of the canyon is probably the worst hole in America, if not in the world. The gloomy black rocks of the Archean formation drive all the spirit out of a man. And the excessive drenching and hard work drive all the strength out of him and leave him in a bad fix indeed. We had to move on or starve. As starving was not on the program, we took our medicine—but I am afraid we did not look pleasant.

Finally we arrived at a peculiarly bad rapid, or rather three rapids, that did not look a bit good. I call them Separation Rapids, as here was where the Howland brothers and Dunn left the party and met their deaths at the hands of treacherous villains destitute of all traces of courage or humanity. Herein find the true story and the cause of the separation—I believe Major Powell calls it desertion. Having been asked to do so by many different people, I now relate the whole story, a story I am willing to swear to at any time, as the facts are as plain in my mind today as incidents of only a few days ago.

As I have said before, when we were about to leave Green River station the signals were arranged thus: Flash flag waved right and left, then down: "Land at once"; right and left: "Slow up"; waved to right: "Follow, but keep to right"; waved to left: "Keep to left"—that is, keep to the right, or left, of pilot boat, the boats to keep one hundred yards apart. O. G. Howland was the draftsman and in charge of the third boat, the "No Name." As already related, too, as we came to the first heavy rapid in Lodore Canyon, near the head of the rapid, I saw the danger and turned in to land. Major Powell was supposed to signal back to the other boats what to do. I doubt very much that he did so. Howland told him after the wreck and rescue at Disaster Falls, when Major Powell angrily demanded of Howland why he had not landed, that he saw no signals to do anything, and was drawn into the rapid before he could see his danger, and then it was too late. I asked about all the men if they saw the signals, and they all said there were none given.

From that date there was more or less rag-chewing on the part of Major Powell, nearly always directed at Howland, until I finally stopped it. At the junction of the Grand and Green there was an outbreak because Howland made a slighting remark about the small amount of supplies gotten at the Uinta

Agency, where there was plenty to be had. Howland would sometimes get a little behind in platting up his survey work, as he had his other work to do the same as the rest of the men. That always brought censure from Major Powell. I remonstrated with him on several occasions, but was informed that "discipline" must be maintained. I told him that we were not a military or naval band, and that I doubted the propriety of such a mode of procedure, as I thought it would lead to much trouble, if not disaster.

My advice was heeded in regard to Howland for a few days, but the trouble was switched off onto Dunn because he had carried into the water the only watch that would run. Dunn kept the barometric notes and had to have the watch, and really should have been a little more careful with it. But it would have done just as well if he had been told to do so in a decent manner. However, the Major evidently wanted to impress his military standing upon Dunn, and proceeded to give him a tongue-lashing that roused his ire to such a pitch that I think that only the fact that the Major had but one arm saved him from a broken head, if nothing worse. Things went on reasonably well for perhaps ten days. Then, after having had a spat with Howland in the forenoon, Major Powell at the noonday camp informed Dunn that he could leave the camp immediately or pay him fifty dollars a month for rations as long as he was with the outfit.

As that little statement raised me to a white heat, I interposed and said that if any one voluntarily wished to leave, he could do so, but that no one could be driven from the outfit. Major Powell informed me that he was talking then, and commanding the expedition. I told him he could talk all he pleased, but that he must cease, then and there, his abuse of Howland and Dunn. Walter Powell tried a bluff and was immediately called to settle, as there was a pretty little sand bar just about long enough for Colt's forty-fours. He replied that he had no arms, and I told him to take his choice of my pistols and choose his distance. He did not accept the proposition. After that little episode, everything was as smooth as with two lovers after their first quarrel and make-up. Major Powell did not run the outfit in the same overbearing manner after that. At a portage or a bad letdown he took his geological hammer and kept out of the way.

Arriving at Separation, or Desertion, Rapid, just as you have a mind to call it, about ten o'clock in the morning, we appeared to be up against it sure. There were two side canyons coming into the Colorado nearly opposite each other, the river not being over fifty yards wide and running like a race horse. The first part of the rapid is caused by the big rocks carried out of the side canyon coming in from the south. The second part of the rapid is formed by rocks from the northside canyon and a granite reef that reaches one-third of the way across the river, making a Z-shaped rapid. We spent the day trying to solve the problem.

Howland and Dunn told me at noon that they did not care about taking any more chances on the river, and proposed to leave and try to make the Mormon settlements north of the river. I did what I could to knock such notions out of their heads, but as I was not sure of my own side of the argument, I fear I did not make the case very strong, certainly not strong enough to dissuade them from their plans. I talked with Major Powell quietly on the subject. He seemed dazed by the proposition confronting us. I then declared that I was going on by the river route, and explained my plans to him how to surmount the difficulty, plans which were carried out next morning. I explained to the boys, the Howlands and Dunn, how we could pass the rapid. In this they agreed with me. But what was below? We knew that we were less than seventy miles from the mouth of the canyon, but had no idea what kind of river we had to encounter below where we were.

As O. G. Howland, who appeared to be the leader of the three, had fully made up his mind to quit, since the rapids had become a holy terror to him, I saw that further talk was useless, and so informed Major Powell. I suggested that we make duplicate copies of our field notes, and give the men the latitude and longitude and a draft of the location of the Mormon settlements as far as we knew. Major Powell and I were up most of the night to get an observation, and he worked out the calculations while I kept a light burning for him with mesquite brush. At daylight we crossed back to the north side of the river and commenced to make the portage, which was accomplished in this way:

All hands took hold and we lifted the boats over the granite reef, about twenty-five feet high. By then taking a bight with the rope around the channels worn in the granite, we managed to let the boats down into a little cove formed below the reef. There was hardly room for two boats to lie, but after getting them there we saw that by striking the main current bows up stream we could make it. If we struck it that way it was my opinion that the current would have a tendency to throw the boats to the opposite side of the river, and thus enable us to avoid the dangerous rock some fifty yards below.

After getting all ready we had a short talk with the boys (the Howlands and Dunn) and urged them to join us. It proved a failure, so we gave them duplicate notes, two watches, and some money. Then Major Powell took his seat in the first or outside boat, while I held it to the bank till all was ready. When the word was given I shoved the boat out as far as I could. The first start was not successful, but the second time away they went. The current took hold of the bow and whirled the boat round in a flash, but shot her nearly across the river, and cleared the dangerous rock and fountain waves below. We watched them till they passed the two falls farther down and saw them turn in and land on a sandy beach a half-mile beyond.

We repeated the same tactics with the other boat and I rounded in beside them. We waited about two hours, fired guns, and motioned for the men (the Howlands and Dunn) to come on, as they could have done by climbing along the cliffs. The last thing we saw of them they were standing on the reef, motioning us to go on, which we finally did. If I remember rightly, Major Powell states it was not as bad as it looked, and that we had run worse. I flatly dispute that statement. At the stage of water we struck, I don't think there would be one chance in a thousand to make it by running the whole rapid.

Before starting I tied the little "Emma Dean" at the head of the rapid. Here we cached two barometers and a bunch of beaver traps. Everything was probably washed into the river at the next flood, as the rise in the river at that point is very great. We saw drift wood in the cracks of the cliffs seventy feet above the river, and we were then in a pretty high stage of water. We ran down a short distance and camped for the night. While we repaired the boats the boys discusssed the conduct and the fate of the three men left above. They all seemed to think the red bellies would surely get them. But I could not believe that the reds would get them, as I had trained Dunn for two years in how to avoid a surprise, and I did not think the red devils would make open attack on three armed men. But I did have some misgiving that they would not escape the double-dyed white devils that infested that part of the country. Grapevine reports convinced me later that that was their fate.

About noon (August 29th) the doors of the Grand Canyon were opened, and we could see all around us, a thing we had not been able to do for many a day unless we climbed the cliff. It was a strange and delightful sensation. For my part, I felt as Dante probably felt as he crowded up from Hades. Paddling along slowly, we acted like schoolboys let out for recess, and "Old Shady" (Walter Powell) made the valley ring with songs. Hall, trying to follow suit, made a failure, and Hawkins told him to shut up or he would drown him. We went into noon camp on the left bank of the river and had a dinner of "sinkers," pure and simple. We saw here the first California quail we had ever seen. There were moccasin and barefoot tracks at our camp, and I thought it a good scheme to get out some reserve cartridges that I had previously stored away in a rubber boot leg, and sealed with piñon pitch. As it happened, we did not need them.

After dinner we ran perhaps three miles. A squaw suddenly bounced out of some mesquite bushes and cut out down the sand at a gait that would have thrown dust in the eyes of Maud S. As she was clothed only in chastity and appeared to have some business down farther, we did not interview her ladyship. A half-mile farther we turned a bend in the river and came to a wikiup with one of Nature's noblemen squatting in front. As we approached the mansion, our Lady Godiva (this one, however, traveled on foot) put in an appearance, evidently to protect her lord. As her first salutation was, "Heap hungry! Towae (tobacco)!" we knew there must be white men in the neighborhood. We tossed the Indians some tobacco and Hawkins threw a cake of fancy-colored soap to one of the papooses. He caught it with the dexterity of a monkey and proceeded to devour it without further ceremony. Running about three miles farther, we came in sight of three men seining a mile below us. With the telescope I saw at once that they were white men. I yanked the colors out of the locker and hoisted them at the bow. We had played our hand out and won against desperate odds! Whether the game was worth the powder or not, I leave to the judgment of others.

Rounding in at the camp of the whites, we found we were at the mouth of the Virgin River, and that the men were some Saints starting a new settlement. They had no horses, so we hired a buck Indian to go up to the Mormon town of St. Thomas after mail. We proceeded to take a rest after catching a lot of fish, which Hawkins baked at once in an outdoor oven. Here comes in a strange case of superstition or some other nonsense. When the fish were done, we offered a good portion to the Indians who stood around us. They would not touch them, but at once rounded up and caught a score of lizards about a foot long. These they ate, not going to the trouble of killing and dressing them. They took them raw as a white man does his oysters. I presume it is a matter of education. The Navajos claim they drove the ancient cliff dwellers in the Colorado River and they turned to fish. Therefore, these Indians will not eat fish of any kind.

Toward evening of the second day the Indian we had sent for mail returned, and with him came Bishop Leithhead of the Mormon Church and Andy Gibbons, a noted scout and guide. They brought a load of melons and other garden truck and two sacks of flour with them. After reading our letters and eating a square meal of "biscuit bread," we commenced formulating some plans for the rescue of the Howland brothers and Dunn. As we described the men and the locality in which they left us, we were listened to in respectful silence. But when Major Powell made the foolish break of telling them the amount of valuables the boys had, I noticed a complete change in the actions of a certain one of the men present. From one with a listless demeanor he instantly changed to a wide-awake, intensely interested listener, and his eyes snapped and burned like a rattlesnake's, particularly when Major Powell told him of an especially valuable chronometer for which he had paid six hundred and fifty dollars.

I believe when Major Powell got up into the Mormon settlements he sent out a party to look for our lost men. I heard about two months afterwards, while at Fort Yuma, California, that they brought in the report that the Howland brothers and Dunn came to an Indian camp, shot an Indian, and ravished and shot three squaws, and that the Indians then collected a force and killed all three of the men. But I am positive I saw some years afterwards the silver watch that I had given Howland. I was with some men in a carousal. One of them had a watch and boasted how he came by it. I tried to get hold of it so as to identify it by a certain screw that I had made and put in myself, but it was spirited away, and I was never afterwards able to get sight of it. Such evidence is not conclusive, but all of it was enough to convince me that the Indians were not at the head of the murder, if they had anything to do with it.

The Powells left us next morning, going with the Bishop Leithhead to Salt Lake City, leaving the necessary instruments with me to finish up the work to connect onto Lieutenant Ives' work, made from the gulf up the river in 1857–58. I collected the stuff and we leisurely pulled out and down the river. Our party was reduced to four, Hawkins, Bradley, Hall, and myself. We drifted down to the old town of Callville, then entirely deserted. We went on and camped that night at the head of Black Canyon. We stopped over at Camp Eldorado, a mining camp where there were about fifty men, and had, of course, to listen to that silly story of James White's navigating the Colorado River on a raft. He was probably some renegade horse thief that had to leave between two days, and very likely struck the Colorado River at the mouth of the Grand Wash.

After taking on some snake medicine, we pulled down to Fort Mojave, Arizona, a two-company post in command of Colonel Stacy, who treated us very kindly. Here we drew four months' supplies and stopped several days. After writing up my journal and getting wind, chart, and weather notes, and other data of interest, we pulled on down to Camp Colorado, where a company of infantry was stationed. Bradley and Hawkins went on down the river while Hall and I were stopping at Mojave. The party was then mighty small, just Hall and myself. We soon passed the other men at Ehrenburg, a small shipping station for Prescott and the adjacent mining country, and then on to Fort Yuma, California, opposite the mouth of the Gila River. After camping there several days to get weather records from the post and take in the sights, we dropped down the river and camped at Bowman Ferry. From there we rowed down through the dismal mud flats of the lower Colorado and camped on an island.

Pulling out from our island camp, we dropped past the squaws and wikiups of the Cocopas, and about noon came to tidewater at the head of the Gulf of California. Not wishing to locate a ranch there, we stayed only two hours. Starting back up the river, we made good progress by using a wagon sheet for a sail. I believe Hall and I are the only men that ever navigated the Green and Colorado rivers for so long a distance—from Green River Station, Wyoming, to tidewater at the head of the Gulf of California.*

As I shared the blankets with Major J. W. Powell for two years, I believe I knew him—perhaps not. He gave very scant credit to any of his men. It was all "Captain Powell and I," when as a matter of fact Captain Walter Powell was about as worthless a piece of furniture as could be found in a day's journey. O. G. Howland was far his superior physically and morally and, God knows, away above him mentally. I have seen it in print that Captain Powell suffered so much in Confederate prisons that it unbalanced his mind. He is still living in Los Angeles, California. [He died March 10, 1915.]

Major Powell paid me but $75.00 for my two years' work. I paid out of my own pocket more than a thousand dollars to make the trip a success. I have seen it in print, too, that the Major spent great sums of his

---

* This was correct up to 1913 when (the Kolb brothers having finished their photographic trip at the Needles in 1912) E. L. Kolb went on to the Gulf, having started from Green River, Wyo., in 1911.—R. B. S.

own money to help the expedition along. May I
kindly ask when and to whom he paid it? But he has
passed over the range, and the Bookkeeper is there.
He was a man of many traits, good, bad, and indifferent. He was vastly over-estimated *as a man,* as so
many others have been. As a scholar and scientist he
was worthy of all praise. His body has gone to the
dust and his soul back to the God who gave it. Let us
cast a mantle of charity over his faults.

I know it is decidedly bad form for me to write as
I have done. My only defense is that in the Report
the Major tries to make a bad impression against the
Howland brothers and Bill Dunn. Bill Dunn was no
saint, but he was, or rather would have been, another
Jim Bludsoe, had the opportunity presented itself.
Hall is also dead—fell in defense of the United States
mail which he was carrying at the time, fighting one
against four, and held it to the last. Two of his assailants were hanged and two are serving life sentences in the Arizona penitentiary. Bradley died some
years ago at San Diego, California, from injuries
received in an accident. Hawkins is still living at
Eden, Graham County, Arizona. I saw him in 1900.
[Hawkins died June 21, 1919.] As for myself, it is
not for me to speak. I have heard it said, "Success is
a virtue and failure is a crime." Nearly all of my companions have passed over the range, and I am still
left behind.

"Ho! Stand to your glasses steady!
'Tis all we have left to prize.
A cup for the dead already—
Hurrah for the next that dies!" *

—JACK SUMNER

* Sumner was the next. He died at Vernal, Utah, July 5, 1907.

CHAPTER IV

MAJOR POWELL'S VERSION REFUTED

WE come finally to a careful consideration of Major Powell's own written explanation of the most dramatic and tragic incident of his 1869 Colorado River voyage. The Major kept a diary or Journal during part of the trip. For the periods May 24th to July 2nd and August 28th to August 30th no journal has ever been discovered. Therefore, to get Major Powell's account of the Separation Rapids story we must turn to the Report, written five years after the events which it purports to record as on the day of their happening:

"*August 27.* This morning, the river takes a more southerly direction. The dip of the rocks is to the north, and we are rapidly running into lower formations. Unless our course changes we shall very soon run again into the granite. This gives us some anxiety. Now and then the river turns to the west, and excites hopes that are soon destroyed by another turn to the south. About nine o'clock, we come to the dreaded rock. It is with no little misgiving that we see the river enter these black, hard walls. At its very entrance we have to make a portage; then we have to let down with lines past some ugly rocks. Then we run a mile or two further, and then the rapids below can be seen.

"About eleven o'clock, we come to a place in the river where it seems much worse than we have yet met in all its course. A little creek comes down from the left. We land first on the right, and clamber up over the granite pinnacles for a mile or two, but can see no way by which we can let down, and to run it would be sure destruction. After dinner we cross to examine it on the left. High above the river we can walk along on the top of the granite, which is broken off at the edge, and set with crags and pinnacles, so that it is very difficult to get a view of the river at all. In my eagerness to reach a point where I can see the roaring fall below, I go too far on the wall, and can neither advance nor retreat. I stand with one foot on a little projected rock, and cling with my hand fixed in a little crevice. Finding I am caught here, suspended 400 feet above the river, into which I should fall if my footing fails, I call for help. The men come and pass me a line, but I cannot let go of the rock long enough to take hold of it. Then they bring two or three of the largest oars. All this takes time which seems very precious to me; but at last they arrive. The blade of one of the oars is pushed into a little crevice in the rock beyond me in such a manner that they can hold me pressed against the wall. Then another is fixed in such a way that I can step on it, and thus I am extricated.

"Still another hour is spent in examining the river

from this side, but no good view of it is obtained, so now we return to the side that was first examined, and the afternoon is spent in clambering among the crags and pinnacles, and carefully scanning the river again. We find that the lateral streams have washed boulders into the river, so as to form a dam, over which the water makes a broken fall of eighteen or twenty feet; then there is a rapid, beset with rocks, for two or three hundred yards while, on the other side, points of the wall project into the river. Then there is a rapid, filled with huge rocks, for one or two hundred yards. At the bottom of it, from the right wall, a great rock projects quite half way across the river. It has a sloping surface extending up stream, and the water, coming down, with all the momentum gained in the fall and rapids above, rolls up this inclined plane many feet, and tumbles over to the left. I decide that it is impossible to let down over the first fall, then run near the right cliff to a point just above the second, where we can pull out into a little chute, and, having run over that in safety, we must pull with all our power across the stream, to avoid the great rock below. On my return to the boat, I announce to the men that we are to run it in the morning. Then we cross the river, and go into camp for the night on some rocks, in the mouth of the little side canyon.

"After supper, Captain Howland asks to have a talk with me. We walk up the little creek a short distance, and I soon find that his object is to remonstrate against my determination to proceed. He thinks that we had better abandon the river here. Talking with him, I learn that his brother, William Dunn, and himself have determined to go no farther in the boats. So we return to camp. Nothing is said to the other men.

"For the last two days our course has not been plotted. I sit down and do this now, for the purpose of finding where we are by dead reckoning. It is a clear night, and I take out the sextant to make observations for latitude, and find that the astronomic determination agrees very nearly with that of the plot—quite as closely as might be expected, from a meridian observation on a planet. In a direct line, we must be about forty-five miles from the mouth of the Rio Virgen. If we can reach that point, we know that there are settlements up that river about twenty miles. This forty-five miles, in a direct line, will probably be eighty or ninety in the meandering line of the river. But then we know that there is comparatively open country for many miles above the mouth of the Virgen, which is our point of destination.

"As soon as I determine all this I spread my plot on the sand, and wake Howland, who is sleeping down by the river, and show him where I suppose we are and where several Mormon settlements are situated.

"We have another short talk about the morrow, and he lies down again; but for me there is no sleep. All night long, I pace up and down a little path, on a few yards of sand beach, along by the river. Is it wise to go on? I go to the boats again, to look at our rations. I feel satisfied that we can get over the danger immediately before us; what there may be below I know not. From our outlook yesterday on the cliffs the canyon seemed to make another great bend to the south, and this, from our experience heretofore, means more and higher granite walls. I am not sure that we can climb out of the canyon here, and, when at the top of the wall, I know enough of the country to be certain that it is a desert of rock and sand, between this and the nearest Mormon town, which, on the most direct line, must be seventy-five miles away. True, the late rains have been favorable to us, should we go out, for the probabilities are that we shall find water still standing in holes, and, at one time, I almost conclude to leave the river. But for years I have been contemplating this trip. To leave the exploration unfinished, to say that there is a part of the canyon which I cannot explore, having already almost accomplished it, is more than I am willing to acknowledge, and I determine to go on.

"I wake my brother, and tell him of Howland's determination, and he promises to stay with me; then I call Hawkins, the cook, and he makes a like promise; then Sumner, and Bradley, and Hall, and they all agree to go on.

"*August 28.* At last daylight comes, and we have breakfast, without a word being said about the future.

The meal is as solemn as a funeral. After breakfast, I ask the three men if they still think it best to leave us. The elder Howland thinks it is, and Dunn agrees with him. The younger Howland tries to persuade them to go on with the party, failing in which he decides to go with his brother.

"Then we cross the river. The small boat is very much disabled, and unseaworthy. With the loss of hands consequent on the departure of the three men, we shall not be able to run all of the boats, so I decide to leave my "Emma Dean."

"Two rifles and a shot gun are given to the men who are going out. I ask them to help themselves to the rations, and take what they think to be a fair share. This they refuse to do, saying they have no fear but that they can get something to eat; but Billy, the cook, has a pan of biscuits prepared for dinner, and these he leaves on a rock.

"Before starting, we take our barometers, fossils, the minerals, and some ammunition from the boat, and leave them on the rocks. We are going over this place as light as possible. The three men help us lift our boats over a rock twenty-five or thirty feet high, and let them down again over the first fall, and now we are all ready to start. The last thing before leaving, I write a letter to my wife, and give it to Howland. Sumner gives him his watch, directing that it be sent to his sister, should he not be heard from again. The records of the expedition have been kept in duplicate. One set of these is given to Howland, and now we are ready. For the last time, they entreat us not to go on, and tell us that it is madness to set out in this place; that we can never get safely through it; and, further, that the river turns again to the south into the granite, and a few miles of such rapids and falls will exhaust our entire stock of rations, and then it will be too late to climb out. Some tears are shed; it is rather a solemn parting; each party thinks the other is taking the dangerous course.

"My old boat left, I go on board of the "Maid of the Canyon." The three men climb a crag that overhangs the river to watch us off. The "Maid of the Canyon" pushes out. We glide rapidly along the foot of the wall, just grazing one great rock, then pull out a little into the chute of the second fall, and plunge over it. The open compartment is filled when we strike the first wave below, but we cut through it, and then the men pull with all their power toward the left wall, and swing clear of the dangerous rock below all right. We are scarcely a minute in running it, and find that, although it looked bad from above, we have passed many places that were worse.

"The other boat follows without more difficulty. We land at the first practicable point below and fire our guns as a signal to the men above that we have come over it safely. Here we remain a couple of hours, hoping that they will take the smaller boat and follow us. We are behind a curve in the canyon and cannot

MAJOR POWELL'S VERSION REFUTED 221

see up to where we left them, and so we wait until their coming seems hopeless, and push on."

For many years we had no written account of this incident other than that of Major Powell himself. Now that we have the two other accounts, that of Jack Sumner and that of William Hawkins, just where lies the truth? Perhaps I am not capable of answering this question, and yet for two reasons I know no one who can better understand its meaning.

First, in this very rapid (my No. 465) I came near losing my life in 1890. I studied it carefully at that time. I reproduce from my note book the account I wrote of the occurrences at this noted rapid on that day in the exact words as written, with the sketch I then made, March 13th, 1890, part before and the balance three hours after we ran the rapid. It was my custom to take the time by my watch and record it in my book at each rapid when entering and coming out. Hence the hours and minutes here given.

March 13th, 1890.

After taking photographs, we start again at 2:45 P. M., and at 2:50 stop to examine Rapid No. 465. This rapid is formed by two large side canyons coming in directly opposite each other, the larger one on the right.

We stop some distance above on right bank, climb over the broken-up cliff, above the side canyon and down across it and on to the high cliff below. The form of the rapid is thus:

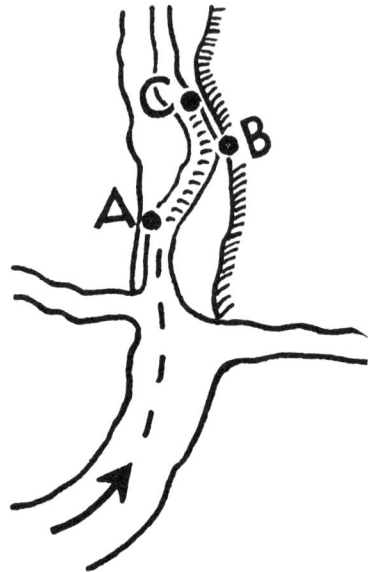

The cliffs are vertical on both sides. On left for 300 to 400 feet, perhaps more, and on right side, 80 to 100 feet.

The water fills the whole channel; a portage of boats or dropping by lines is impossible (of boats as large and heavy as ours were then). The current draws in at head of rapid over huge rocks on both sides and drops very suddenly over first cliff, just as it goes over second fall, and the rebounding waters from the cliff, the rushing water from the right, and the fall, make a torrent of splashing, curling, jumping waves, broken into one mass of yellow, dirty foam, such as our boats never went into before.

After the current leaves this point (A) it all rushes on to the cliff at "B," and from "B" to "C," along the cliff, is one immense *boil* and then the current turns sharp around the point "C" at which is a sharp, projecting rock.

After examining carefully, we conclude this rapid must be run; there is no other way. There is no hesitation on the part of anyone. We start into the rapid in good spirits and good shape at 3:07 P. M."

Second, in my work on the Colorado, I occupied at different times two distinct positions. From May to July, '89, I was in a subordinate position under the command of another, and from that standpoint I understood the position of the men. And again, from December, '89, to May, '90, I was the commander of the expedition, and from that standpoint, with my former experiences, I think I understood the commander, and also the men. One of my men left me and I discharged three others while I was on the river. I hardly think there is another person who ever occupied these two positions in the exploration of the Colorado.

Since this unfortunate incident of the Howland brothers and Dunn has been somewhat discussed already by F. S. Dellenbaugh (*The Romance of the Colorado River,* pages 226 and 227) it cannot be out of place for me to continue that discussion and bring out more facts in the case. Considering all the evidence and testimony, I arrive at certain conclusions.

No one who ever knew Major Powell would for one moment attribute to him one particle of fear of the rapids, or any want of determination to complete the journey. On the other hand, the whole tenor of the Major's account is that the Howland brothers and and Dunn left the party on account of the severity of that one rapid, and for no other reason. As their companions put it, the Major accuses these men of personal cowardice. This I consider entirely unjust. Those three men were as brave as any upon the river. Very seldom do frontiersmen such as they desert their comrades in the face of danger, as we who have known and lived with them most certainly know.

Furthermore, Major Powell places the whole burden of the separation upon the elder Howland, only incidentally mentioning Dunn. It is my carefully studied opinion, formed from the facts as given me by Sumner and Hawkins, that the reason for their leaving was that Dunn had been *ordered to leave* a short time previously. This he had promised to do as soon as he could. He chose this particular point at which to act upon his peremptory discharge. The elder Howland was only the spokesman for his friend Dunn. Sumner says Howland seemed to be the leader, that is, the elder and the spokesman, as Hawkins says Sumner was on another occasion.

That the severity of that particular rapid had its influence upon their choosing that particular side canyon as a place of departure is undoubtedly true. The apology has been made by Dellenbaugh in the discussion above referred to that "Powell may have dis-

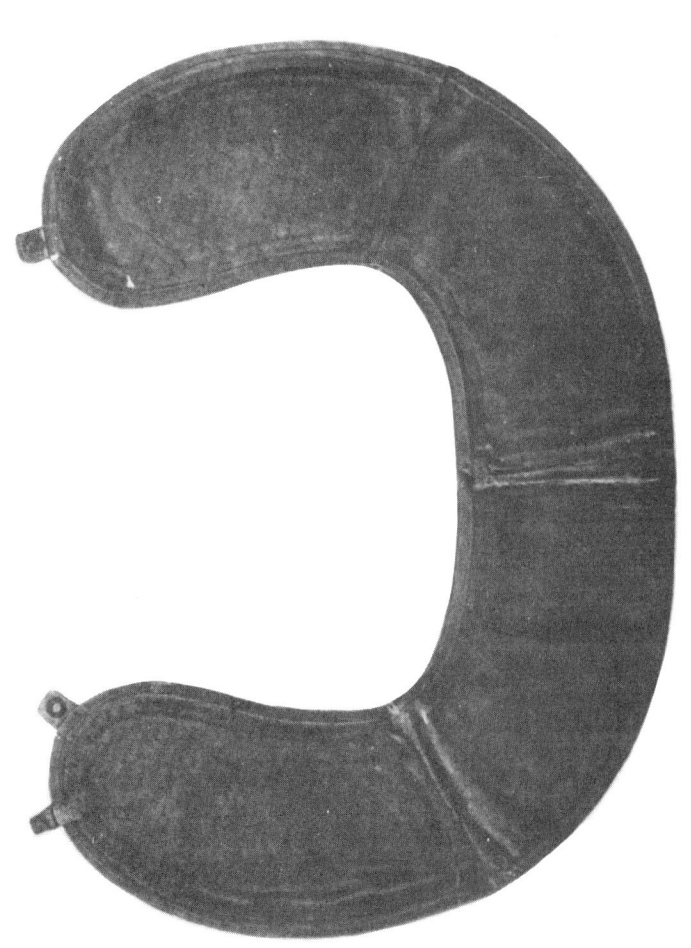

LIFE PRESERVER USED BY MAJOR POWELL ON THE FIRST EXPEDITION
Now in the Smithsonian, its existence was denied by a studied silence on the part of the Major for many years

covered Howland *persona non grata,* but had this been as serious as some have said, Howland would not have waited, it seems to me, till they came to a particularly bad-looking place to take his departure." (*The Romance of the Colorado River,* p. 227.) But this, it seems to me, is far from the mark. It was not Howland but Dunn who left. The elder Howland, with additional reason of his own, stuck by his friend, and the younger Howland stuck by his brother. It was Dunn who was ordered to leave only a comparatively short distance above this place and at a point where it was impossible for him to climb out. He merely waited to leave by the first promising side canyon.

That particular rapid undoubtedly had its bearing upon the particular day and time of departure, but it was not the real reason. I agree with Sumner and absolutely disagree with Powell when the latter says, "We have passed many places that were worse." That rapid, though it has been run successfully several times since, is in my opinion the most dangerous one to navigate upon the whole river, especially at certain stages of water. In substantiation of this (never mind what Powell wrote in his alleged daily journal) we have the testimony of one of Powell's men as to what Powell really thought of it. Dellenbaugh, giving the reason why Powell ordered the second expedition abandoned at the mouth of Kanab Canyon, September 7th, 1872, writes: ". . . but Powell announced that he had decided to end the river work at this point, on account of the extreme high water, which would render impassable the rapid where the Howlands and Dunn had left." (*The Romance of the Colorado River,* page 341.)

The high water had not deterred him from running the "Sockdolager" and the other great rapids above. Yet in 1872 Powell feared the consequences of "Separation Rapid." What he wrote in his report in 1874 was another matter. What was put in his diary in 1869 on those memorable days actually "written on the spot," is this and nothing more:

"Aug. 26. Found Indian Camp today, gardens. Good River, 35 miles. Camp 42.
"Aug. 27. Run 12 miles, at noon came to bad rapid. Spent afternoon in exploration. Camp on left bank in Gulch, 43.
"Aug. 28. Boys left us. Ran rapid. Bradley boat. Make camp on left bank. Camp 44."

And here ends the original journal. There is not another entry after August 28th, 1869.

Naturally the rapid affected the whole party. After months of hardship, near starvation, and nerve-racking work on the river, this is not surprising. For the reasons already stated I am inclined to give full credence to Hawkins' story that at one time it was actually determined that the whole party should abandon the river at that point. Major Powell himself wrote in

MAJOR POWELL'S VERSION REFUTED 227

1874 in his Report: "At one time I almost conclude to leave the river." Had this been "written on the spot" in 1869, I am satisfied it would have been, "At one time I conclude to leave the river." This in no way reflects upon Major Powell's bravery or his desire or determination to go on, but without his men that would have been impossible. Sumner acknowledges that when he was trying to dissuade Howland and Dunn from going out to the Mormon settlements, he was not sure of his own position. He says, "I did what I could to knock such notions out of their heads, but I was not sure of my side of the argument."

When the other boys, the youngest members of the party, refused to follow the Major out by land, the tide was turned. However, all this in no way essentially alters the fact that the three men who left the river had another and much more potent reason for cutting loose from their companions in peril.

I very much doubt the whole narrative, too, as to sequence of events and details recorded by Major Powell in the Report. Very properly it may be asked why I assume to so discredit Major Powell's statements, though accepting those of Sumner and Hawkins. For this reason: every statement made by Sumner and Hawkins to me, verbally or in writing, with regard to the physical conditions of the canyons and the river I have found invariably correct. I cannot say as much for the Major, unfortunately. In this light let us examine Major Powell's narrative of these two days, though not exactly in the order of his writing.

The account of his examination of the rapid from the left side, where he is caught on the cliff and rescued by the use of oars is beautifully and dramatically written. Still, what proof have we that it ever occurred at that particular rapid or on that particular day? Some of Major Powell's experiences "written on the spot," you will remember, never occurred until years after '69. Something like this may have occurred somewhere else and been transferred in time and place to suit this particular occasion. Even if it did happen as related, it was a useless risking of his life, for he records: "We land first on the right and clamber up over granite pinnacles for a mile or two, but can see no way by which we can let down, and to run it would be sure destruction," showing that on the first examination, at least, he decided it was impossible to run that rapid.

But in which direction did he clamber "for a mile or two"? Certainly not *up* the river, away from the rapid. Certainly not at right angles from the river away from the rapid, nor vertically up over canyon walls. Not at all. They walked *down* the river, along the granite, some little distance above and along the river, where they saw the whole stream and the rapid so easily and so well that Powell at once decided that

MAJOR POWELL'S VERSION REFUTED 229

there was "no way to let down, and to run it would be sure destruction."

Such an examination could be easily made on the right side, where he says it was made, and such a clamber could easily be taken on that side. When there in 1890, I took that identical walk, though not for two miles, for that would have been a waste of time. In fact, it would have carried me more than a mile and a half below the rapid. For this reason I doubt absolutely the veracity of Powell's phrase "for a mile or two." I made just such an examination, and at once decided that we must run the rapid or leave the canyon. We ran it within the next five minutes afterwards.

In 1908 E. R. Monett took the same walk, as walks go in the Grand Canyon, and carried on his back for safety a lot of glass photographic plates while Charles Russell ran the rapid in his boat. Why, then, the story of the dangerous climb along the wall so high above the river on the left side? Did such a climb ever occur? I don't know. But photographs of the spot show no necessity for such a climb on the left side. In the afternoon Major Powell made a second careful examination along the right side, where, he says, he saw the whole situation so well that he decided it could be run. It was after this, according to the Major's Report, sometime during the night, that he "almost" concluded to leave the river. It was after breakfast the next morning, the 28th, that he announced, according to Hawkins' account, "Well, Billy, we have concluded to abandon the river for the present."

The three men having made their final decision to leave the party, I can well understand, under the circumstances, the terrors of that night for every one, and why it was that for the Major there was no sleep. And what a terrible thing it was for him to feel that he must leave the exploration unfinished! After it had been decided, however reluctantly, that the whole party should abandon the river, a fourth change of plan came from the Major to Hawkins, or from Hawkins to the Major, as you choose. But for other reasons this did not alter the determination of the three men to get out to the settlements by way of the side canyon.

Having run the rapid, Major Powell writes: "We land at the first practicable point below, and fire our guns as a signal to the men above that we have come over in safety. Here we remain a couple of hours, hoping that they will take the smaller boat and follow us. *We are behind a curve in the canyon* and cannot see up to where we left them, and so we wait until their coming seems hopeless, and push on."

Jack Sumner states: "We waited for about two hours, fired our guns, and *motioned the men* (the Howlands and Dunn) *to come on, as they could have done by climbing along the cliffs.* The last thing we saw of them, they were standing on the reef *motioning*

MAJOR POWELL'S VERSION REFUTED 231

us to go on, which we did." *

And I add my testimony to this statement of Sumner's that anywhere at the edge of the river at "the first practicable landing," and at several good landings farther down (unless standing behind a boulder), one can see *up that straight stretch of the river over* the whole rapid, and might easily see a man standing on the reef, as Sumner says, and as photographs of the spot show.

If there had been no other reason for the men to leave but fear of the rapid, the Major or his men could have walked back to them over the granite trail over which the Major says he had already passed four times, and then walked with them (or even carried them, if necessary) down to the landing. Then the entire party could have proceeded in the two boats.

The whole story of those two memorable days of August 27th and 28th, 1869, as set down in the Report of 1875, is beautifully done—as a piece of narrative writing. As *history,* however, it is sadly deficient and misleading. As has already been pointed out with sufficient emphasis, the fateful "curve" in the canyon which Major Powell declares cut him off from view of the men left behind actually existed only in his mind

---

* In the article already referred to written by Major Powell and republished in 1870 in Dr. Bell's book, the Major says: "The men that were left sat on the cliffs and watched us go safely over, so we went into camp and waited two hours," etc. That was undoubtedly true, so that if the Major could see those men, and know that they were able to see the whole running of the rapid, why was that "curve" in the canyon invented years after at that particular spot? The answer has already been supplied.—R. B. S.

when he was writing his official Report in 1874. His invention of that extremely convenient if quite imaginary curve furnishes the climax, so to speak, of a deliberate effort to obscure the conditions which caused O. G. Howland, Seneca Howland, and William Dunn to go from the expedition to their deaths.

Considering the true state of affairs, there seems to be no real reason why the plate on the Powell Monument in Grand Canyon National Park should not be replaced with another whose inscription includes the *full* roster of the 1869 expedition. So long as this remains undone there is perpetuated a gross injustice to the names of three brave men—perpetuated by the same government which (except for Major Powell's regrettable misrepresentation and evasion) would have been glad to do them honor when on May 20th, 1918, there was dedicated this memorial from the Congress of the United States to the first conquerors of the Colorado.

APPENDIX

Commentary on

PART ONE
JAMES WHITE'S RAFT JOURNEY OF 1867
By Otis R. Marston

The Legendary Conception

"A legendary conception of the Far Western Frontier has been established in the minds of the American people," Harvey L. Carter once wrote. "The effect has been, not to cause Americans to become interested in the study of history, but to cause them to become so attached to the legend that they actually offer mental resistance to the efforts of historians to replace legend with fact. . . . Historians, themselves, are not immune to the effects of the legend."

To establish the legend that a Colorado resident had been the first to make a water transit of the Grand Canyon was the purpose of a U.S. Senate Document in 1917.

The letter of February 8, 1917, written to White's daughter by Thomas Fulton Dawson was part of a barrage of letters directing her preparation of a report signed by White which Dawson managed to have printed as Senate Document No. 42, *The Grand Canyon,* 1917, with the caption on page 39 reading "White's Own Story."

Dawson's position with the Senate would give the illusion of authority, but in the preparation of this work he never met White and made no effort to study any of the character of the Colorado River system in its gorges.

Writing White's daughter, Esther, February 8, 1917, on stationery of the Office of the Secretary, United States Senate, Dawson explained:

> I am writing this note with a view to impressing upon you the importance of getting as definite a statement as you can from Mr. White as to his being on the Grand river. If he is in position to give it, the statement should be made over his own signature and largely dictated by himself, and it should be made independent of this letter. Let him begin by saying that he understands that the charge has been made that he did not reach the Colorado river via the Grand river, etc., and say nothing about receiving a letter from me.

Not waiting for a reply, Dawson supplemented with a letter dated February 10:

> I fear I did not make sufficiently plain my reasons for wanting it made clear that the Baker-White party was really on Grand river in the letter I wrote you the other day. This is the one principal one: Grand river is far north of the San Juan, and it would be much more difficult to get from Grand river over land to the lower Colorado than to get there from the San Juan. In fact it would not be so difficult to get there from the San Juan, while it would be almost out of the question to make the trip from the region of the Grand. You will remember that I say in my article that it would be almost as much of a fete (sic) to go

overland as by water. That statement is predicated upon the theory that the party was on the Grand when Baker was killed. So, bring that fact out as clearly and strongly as possible over your father's signature.

On March 8, 1917, Dawson wrote again marking the letter "Confidential" and requesting its return. It was sent back to him April 13 but is missing from the otherwise complete file of the letter exchange now in the State Historical Society of Colorado.

Dawson sent the first copy of the Senate Document to Stanton and this extended the interchange of letters between them. A copy was sent to Frederick S. Dellenbaugh and his letter comments were also extensive. News reports added to the confusion. Some basic interpretations can be obtained by reference to the file of Dawson-White family letters in the State Historical Society of Colorado.

The Record Before White

The Colorado River port of Callville where James White ended his raft journey September 7, 1867, was located on the right bank about one mile below the mouth of Boulder Canyon in the State of Nevada.

Callville had been started in December, 1864, conceived by the Morman leaders as a port of entry for freight brought upriver by paddle wheel steamers after receiving it from ocean

236

vessels near the head of the Gulf of California. Church immigration from Europe was also to use this route after crossing at the Isthmus of Panama. The steamers *Esmeralda* and *Nina Tilden* were able to make several trips to Callville but were handicapped by rocks and sandbars so the effort was abandoned. The chief hazard to the development of this transportation was the completion of the transcontinental railroad and Callville was abandoned in 1869. About 1937 the ruins were submerged by the rising waters of Lake Mead back of Hoover Dam.

At the time of White's arrival at Callville, the river course above that point and below the head of the Grand Canyon was virtually unknown. George C. Yount and James Ohio Pattie, in the spring of 1827, reached the river at Mile 246 (measured downriver from Mile 0 at Lees Ferry, Arizona) via Spencer Canyon and left by the same route. The *San Diego Union* of June 20, 1857, printed a letter of famed guide and trapper, Joaquin Antoine Jacques Leroux, telling of his cruise with canoes in 1837 from the mouth of the Virgen to the Gulf of California. The first twenty miles of his course would have been over the unknown part of the river above the site of Callville.

On April 5, 1858, Lt. Joseph Christmas Ives led an overland party to the mouth of Diamond Creek, Mile 225.7, and they spent the day studying the area. F. W. Egloffstein made a sketch which is highly imaginative.

Late in 1862 Jacob Hamblin, using a small skiff, ferried twenty men across the Colorado at the mouth of Grand Wash, Mile 284.5. In 1863 he used the same craft to cross five miles above Grand Wash, at what is now known as Pierce Ferry.

The San Francisco *Daily Alta California,* May 14, 1864, printed a letter from "Cal. Volunteer," probably Alonzo E. Davis from Fort Mohave, "Messrs. Butterfield and Perry (Ferry?), from LaPaz, passed by here a few days ago, on their way up the river, on an exploring tour. They are making their way in a small boat. . . . It is their intention to go up two hundred and fifty or three hundred miles." On July 4, 1864, the same source reported, "They went up some 260 miles, when they lost nearly all their provisions by the capsizing of their boat, and, of course, were obliged to return. . . . Mr. Ferry says the river is easily navigable, at least in high water, to Black Canyon, about 180 miles above here."

Lt. George M. Wheeler interviewed Octavius Decatur Gass in 1869 at Las Vegas Ranch (located at the intersection of Las Vegas Boulevard and Washington Street in present-day Las Vegas) and learned that Gass and an Indian had joined Butterfield and Ferry in their cruise into the lower end of the Grand Canyon. During Wheeler's cruise in 1871 from Fort Mohave to Diamond Creek, he discovered a monument on October 9 near Mile 264, right bank, and a memorandum was cached in it bearing the signatures of four men. In 1920, Geological Survey topographic engineer, Roland Whitman Burchard, found a monument at Mile 260.3, left bank, and entered it on his maps as Cass (sic) Monument. Gass assured Wheeler that point was the head of navigation, 82.8 miles upriver from Callville.

The *Deseret News,* July 3, 1867, reported the launching of a skiff, 16 feet long, at the mouth of the Grand Wash on April 15, 1867. Henry W. Miller and Jess W. Crosby each handled a pair of oars while Jacob Hamblin took the helm. After pulling a mile and a half upstream, they started down and estimated their cruising speed at five miles per hour. Having cruised through several rapids, they approached the Grand Rapids, whose roar was heard for over a mile before reaching them. They estimated the fall as six to eight feet in twenty rods and the lining of the skiff proved difficult. They judged this would be the head of navigation on the Colorado. Camp was near the Temple which they named Tower Rock. Arrival was at Callville next day and they were confident that no white man had cruised this 60.6 miles before them.

The flow of the river is an important factor controlling navigability. Peak flows on the Colorado River had always occurred in the period from May to early July during the high snow melt runoff. The maximum recorded flood in the Grand Canyon was July 8, 1884, at 300,000 cubic feet per second. Flows of less than 1,000 cubic feet per second were not a rarity. The median peak was about 80,000 cubic feet per second and it could be expected in mid-June. By August 15 the median flow dropped below 10,000 cubic feet per second where it remained through the winter.

Records indicate unusually high flows during the early summer of 1867.

The *Daily Alta California* of July 10, 1867, reported, "21 & 22 June 1867. Wm. H. Hardy rode the steamer *Cocopah* from Hardyville to El Dorado Canon and back on those two days. . . . His report included the following statement: "The colorado at the present time is running with full banks, and is quite as high as ever known by white settlers. . . .' "

An Army inspector, Jones, was at El Dorado Canyon July 2, 1867, and reported on the status of the government cattle supposed to be on Cottonwood Island: "For weeks high water has cut off all communication with the island, and there is no telling when the command at

Camp Eldorado will again be furnished with fresh beef."

The *San Bernardino Guardian* of August 24, 1867, reported: "The Colorado began to fall July 10, when numerous crops were put in, but it is feared too late to amount to much."

The *Arizona Miner,* of Prescott, stated July 13, 1867: "The Colorado is higher than it has been for many years, and at some of the ferries it is impossible to cross."

Among the Mormons, the soldiers, and the barge crew who witnessed White's arrival at Callville September 7, only James Ferry had any direct experience with the river above Callville. Octavius D. Gass was at the meeting of the Arizona Legislature. The limited communication from White, the only witness, and the lack of detailed knowledge of river conditions above Callville invited wild speculation in determining the point of embarkation of the raft cruise. The guessing continued over the years with more than a dozen locations being selected.

The Record After Powell

The Brown-Stanton cruise left Lees Ferry July 9, 1889, riding about 15,000 cubic feet per second of flow. After losing Brown July 10, and two other men on July 15, they left the river at Paradise Canyon at Mile 31.5. Stanton, with better boats and life preservers, departed from Lees Ferry December 28, 1889, and it was March 17 before they cleared the Grand Wash Cliffs. Accidents, photography, and surveys had slowed progess.

William Hiram Edwards had extensive river experience compared to others of his time. He was in the Stanton crew of 1889-1890 cruising from North Wash to the Gulf; with the Best Expedition whose river travel was from Green River, Utah, to Lees Ferry in 1891; the leader of two cruises of the Steamer *Major Powell* which cruised twice in 1893 from near the mouth of the San Rafael, on the Green River, to Spanish Bottom at the head of Cataract Canyon and return. In 1896 Edwards worked for months at the Bennett Amalgamator, below Dellenbaugh Butte on the Green River, which provided the opportunity to visit with the Flavell-Montos and the Galloway-Richmond crews which made transits of all the canyons. In a letter to Stanton dated April 1, 1907, Edwards reported:

> The April Outing magazine has an article on that man White who claims to have gone through the canyon in 1868. The last I knew of the man he lived at Trinidad and I met him about 1894 and had quite a long talk with him. From the story he told me I do not believe he ever went through the canyon. You will note in this Outing article he claims to have floated from somewhere near the present town of Moab, Utah to Colville in fourteen days. The way a raft would drift around in the eddies, I do not believe he could float through there in four-

teen weeks. According to his story Cataract Canyon was all smooth sailing and we know better than that.

Edwards supplied Stanton with White's address in Trinidad, Colorado, and an interview there was arranged in White's home the evening of September 23, 1907.

Nathaniel T. Galloway and William Chesley Richmond, in two skiffs, departed Lees Ferry on low water January 12, 1897, and reached the Virgin River February 3. As leader of the Galloway-Stone party some thirteen years later, the trapper departed Lees Ferry October 28, 1909, and on November 16 the ruins of Callville were behind them. At the time, Galloway was the only man who had made two water transits of the Grand Canyon and both had been on moderate or low river flows. In his journal, printed in his *Canyon Country,* 1932, Julius Frederick Stone commented:

> ... in 1867 James White was taken from a raft after having, as he claimed, passed (on that precarious platform!) through all the canyons from a point somewhere above the junction of the Green and the Colorado in thirteen days — a claim so manifestly incredible as not to win the belief of anyone who has passed through either the Marble or the Grand Canyon.

On April 2, 1917, in a letter to Stanton, Stone was more emphatic:

> It is not only unbelievable but physically impossible that White should have made the trip from above the head of the Colorado to Old Callville on a raft in

fourteen days when it took us over a month with the best possible equipment plus a very considerable experience. It beats all how difficult it is to kill a lie after it has been repeated a sufficient number of times.

After completion of his cruise from Wyoming to the Gulf in 1911-1912, Ellsworth Kolb visited White in Trinidad in 1914. In a letter of December 27, 1922, Kolb wrote to Lewis Ransome Freeman:

> I would like to know the truth about White. I talked with him a few years before he died, but he was so childish it was impossible to make head or tail of his story. He may have gone through the Canyon but he certainly never drifted through in low water in two weeks as he said. He told me he did not travel at night and did not think it was so very bad except for a couple of falls. He also said he could touch both walls of the side canyon he called it the Grand River, sometimes — with the pole he carried on his raft. Others think he meant the San Juan. I asked him why he didn't leave the river in the quiet stretches. He replied he feared the Indians and had no shoes.

Emery Kolb, Eugene Clyde LaRue, and Lewis Ransome Freeman were members of the Geological Survey crew, under the leadership of Claude Hale Birdseye, which embarked at Lees Ferry August 1, 1923, to survey the course of the Colorado River through Marble and Grand canyons. Freeman had gained some Colorado River experience, largely on flat water, and his book *The Colorado River,* had appeared in print a month previously. On page 195, he writes of White's voyage:

> ... I know enough of the stretch between the Grand Wash and the mouth of the Virgin fully to agree with Stanton in identifying it with the portion of the

Colorado most nearly fitting White's description both as to rapids and character and colour of the walls. He could have drifted down here in the sort of rafts he describes, and he could easily have encountered all the trouble he recounts in doing it. . . . Not one man in a thousand — not one in a hundred thousand of White's type — can carry an exact record of river travel in his head. . . . Still less could White have kept a mental record of a voyage which was started in panic, and continued to be more or less of a panic all the way.

Freeman's experiences as an oarsman during the cruise of the Grand Canyon in 1923 added comment in his book *Down the Grand Canyon*, 1924:

There proved to be nothing at the mouth of the Little Colorado resembling even remotely the great whirlpool located there by James White in his lurid account of a raft voyage through the Grand Canyon in 1867, two years ahead of Powell. Indeed, from first to last the observations of the survey were calculated to discredit every statement of a sensational yarn which, even as late as 1917, was considered worth recording in a special Senatorial document. As I have already written at some length in an endeavour to give a fair consideration of White's wholly impossible claims, I mention the matter here merely to record that the intimate knowledge of the Grand Canyon gained in the course of our recent voyage makes the averred adventure of the Colorado prospector appear more utterly absurd than ever.

Men experienced in the rapids of the Grand Canyon through 1923, when they have expressed opinions whether White cruised any of that major course, have agreed that it was impossible.

A contrary view appears in the July 14, 1940,

entry in the diary of Barry Morris Goldwater, a private printing of his *A Journey Down the Green and Colorado Rivers 1940*. In his beautiful book, *Delightful Journey Down the Green & Colorado Rivers* published in 1970 by Arizona Historical Foundation, Tempe, Arizona, the discussion regarding White is shifted to a chapter headed "Afterword." It states:

> The discussion often arose on our trip ... concerning the identity of the first man to go through the canyons of the Colorado, particularly through the Grand Canyon. The discussion usually dissolved into an argument, with those of the White school aligned on one side and those of the Powell school on the other. The former are considerably in the minority. ... As for me, I place confidence in White's story and believe that he made the journey he claims to have made. My friend Dock Marston disagrees.
>
> James White's story, *The Grand Canyon,* was printed by the Government Printing Office in 1917, under Senate Resolution No. 79. You may read his account and draw your own conclusions. To those who say that such a voyage could not be made on a raft, I answer that men in desperate circumstances have accomplished more dangerous feats than running Colorado River rapids on a raft, although I would never, in my weakest moments, venture such a trip. To those who say that White is wrong on certain points, I say to imagine yourself starved, cold, and scared as hell in the middle of the Colorado River on a raft, and then ask yourself whether you would give a tinker's damn about the scenery or details of it. I repeat that nothing yet has been brought forward to make me accept anyone other than White as the first through here.
>
> White's journey contributed nothing to the knowledge of the river, proving only that the river could be traversed.

The following records aid in avoiding the legendary conception in the White study:

The diary of Eugene Clyde LaRue who, with five others, two skiffs and a canoe, started from Grapevine Wash at Mile 279.3 on low water September 25, 1924, and arrived at Callville September 29. A small motor enhanced the speed. Thirty rapids were tallied. Twice the canoe went through upside down. At Hualpai Rapid all dunnage was portaged and the craft were lined.

June 24 to 30, 1946, Harry Leroy Aleson attempted to prove that James White made the cruise as he claimed. With Georgie White, Aleson arrived at the mouth of Parashont Wash, Mile 198.3 in Grand Canyon, and carried a one-man rubber air float, four life preservers, and some limited supplies. Juniper and cottonwood logs were tied with a bridge timber with some light line to make a raft, but four hours of struggle failed to get it clear of the eddy and into the current.

The two embarked on the inflated air float and they were rolled into the river in less than a mile. Caution demanded the portaging of a number of rapids and, on June 26, at the mouth of Fall Canyon, Mile 211.6, a collection of drift invited the construction of a second log raft. "It would not take the current. We marked it No. 2 and abandoned it in the backwater." The record appears in print in the *Boulder City News,* July 2, 1946, and the *Southern Sierran,* August, 1946. The experiment was of value but was faulted by the river flow between 42,200 and 34,600 cubic feet per second.

The speed of a low water run of Grand Canyon is exhibited in the record of seven-foot plastic Sportyaks embarking at Lees Ferry August 5, 1963, and clearing the Grand Wash Cliffs August 31. The flow was limited due to construction of Glen Canyon Dam and varied from 1430 to 2670 cubic feet per second. Progress demanded numerous linings and portages and the upper half of the Canyon required constant work with the oars due to lack of current. A photographic record by William Belknap, Jr. appeared in *Argosy,* May, 1964.

Stanton's Conclusion

In developing the background for his interview with White, Stanton failed to find the William J. Beggs news item published in the *Arizona Miner* of Prescott, September 14, 1867, ten days earlier than the E. B. Grandin item of September 24 in the *Daily Alta California,* which Stanton called ". . . undoubtedly the first published account of White's journey." Nor did he find the reprints of the Beggs account in *Deseret News* January 27, 1868, and in *Lippincott's Magazine* December, 1868.

A public stenographer, Roy L. Lappin, was hired to make notes and prepare a typescript. The first part of the interview gave White the opportunity to tell his story freely. The second part was marred by Stanton giving extensive recitations of some of his limited knowledge of the river course. Lappin's typescript proved to be sketchy but not inaccurate.

During the Stanton interview, White described the failure to establish placer values in the area where Silverton, Colorado, is now located. They traveled south and west to the San Juan River which they reached about the first of August at a point thought to have been twenty-five miles from its mouth. White had no information to determine this distance and the probable crossing was over one hundred miles from the mouth of the San Juan. White was positive that was the only crossing they made of the San Juan.

If the trio then continued southwest as White told Stanton they did, this would have placed them in the Little Colorado drainage. Here numerous dry washes lead down to the bed of that river. The search for water which took Strole up canyon and Baker down would have been typical of the Little Colorado when it was not in flood. When the two water seekers returned to camp, White reported having seen an Indian but his companions judged he was crazy and only imagined seeing one. Camp that night, August 3, was at the bottom of the narrow gorge.

White's presentation has established that he had not embarked on the river in the area of either the mouth of the Green River or at the mouth of the Little Colorado. Directly below the mouth of the latter stream is a rapid of moderate intensity. About thirteen and one-half miles up the Little Colorado from its mouth are the sizable Blue Springs which supply a perennial flow of salt water. In White's letter to Thomas F. Dawson, dated April 20, 1917, he emphatically stated he "did not travel down any small stream before reaching the Colorado River."

If Stanton's conclusion is correct that White and Strole embarked below the Grand Wash Cliffs, the probable point of the shooting of Baker was close to the crossings of the Little Colorado, which had extensive use from those traveling the trails north of the San Francisco Mountains. The Navaho and the Walapai traded extensively. Father Garces was through the area in 1776. In 1854 Aubrey used the trail. After leaving the river in 1858, Joseph C. Ives used the trail to reach the Navaho country. In 1862 and 1863, Jacob Hamblin crossed at two points directly below the Grand Canyon en route to Oraibi. In the fall of 1864, William H. Hardy claimed he started eight prospectors from Hardyville on a three-month investigation. Hardy compiled and mapped their notations of the features of the route to the Little Colorado and gave it to Richard Gird.

The date of Baker's death was probably August 4. With few exceptions the date of arrival of White at the end of his cruise was accepted as September 7. It is also widely accepted that the date of embarkation was August 24 but the authority for this selection must be accepted as weak. The twenty days between the death of Baker and the embarkation of White and Strole on the river would have provided ample time for them to travel from a crossing of the Little Colorado River, near the present location of Cameron, to areas below the Grand Canyon such as Grapevine Wash, an air distance of about one hundred and fifty miles.

Stanton closed his interview of September 23, 1907, with the statement to White: "The fact is that from the time you struck the San Juan River at the mouth of the Mancos River, *you were lost*. You did not know where you were, or where you went." White replied: "Maybe I was!"

The numerous excursions into legendary conceptions provide entertainment, but careful analysis fails to reveal any reason to reject White's raft journey as starting at Grapevine Wash and ending at Callville. No presentation has fixed the date of the embarkation. Much serious study remains to be done.

OTIS R. MARSTON

San Francisco
March 1979

Otis Marston, known as "Dock" to all Colorado River buffs, was for many years the foremost authority on river running in the Grand Canyon. A native Californian, he lived in the San Francisco area, and was at various times an engineer (two degrees), naval officer (submarines), stockbroker, teacher, lecturer, and writer.

Dock "ran the last rapid" August 30, 1979.

Commentary on
PART TWO
THE AFFAIR AT SEPARATION RAPIDS
By Martin J. Anderson

Time has a way of surfacing records which change opinions. There are several statements by Robert B. Stanton, Julius F. Stone, and James M. Chalfant which need correction.

The main objective of the Sumner and Hawkins accounts was to exonerate the Howlands and Dunn from the accusation of being "cowards," "deserters," and mutineers." Stanton, Stone, and Chalfant erroneously blame Powell for putting these labels on the three. There is no record that Powell ever questioned their bravery and courage. It was Frederick S. Dellenbaugh's opinion, not Powell's. In his first book, *Romance of the Colorado River,* 1902, he branded the men deserters. Later, in *A Canyon Voyage,* 1908, he carries over this opinion and adds, "I never heard Major Powell say a word in condemnation of any of these men; on the contrary he always spoke of them affectionately."

The greatest thrill in Dellenbaugh's life was his association in 1871-1873 with the Powell Survey. He was a teenager at the time, and never outgrew his adolescent hero worship of Powell.

Once he formed an opinion neither heaven, earth, or truth could change it. After his books were published he became the self-appointed historian of the Powell voyages.

It was Dellenbaugh who was responsible for the omission of the names of the Howland brothers and William Dunn from the Powell Monument at the South Rim of Grand Canyon. The memorial was planned at the International Geological Congress in 1904, two years after Powell's death. Its purpose was to honor Powell's service as the Director of the U.S. Geological Survey, a position he held from March, 1881, to June, 1894. In 1904 Congress appropriated five thousand dollars for the structure. The original plan was for a Roman chair facing the canyon. Later an altar with a plaque was substituted. Dellenbaugh was consulted throughout its planning, and it was he who suggested the wording on the plaque and the omission of the three men. He boasted of this involvement, and continued to call the three men cowards and deserters.

It was Dellenbaugh who first urged Sumner to write an account of the Powell voyage. After Sumner read Dellenbaugh's *Romance of the Colorado,* a short correspondence developed. Until then Sumner was reluctant to give details of the 1869 voyage, as evident in his meeting with Stanton (page 104). Finally persuaded, he wrote

Dellenbaugh, "I have commenced my version of the so called Powell exploring expedition which will differ from Powell's Report and the reason for the Howland Boys and Dunn left us. . . ." Since Sumner's version was not complimentary to Powell, it appears Dellenbaugh did not pursue the matter. There is no record that he ever publicly admitted having this correspondence with Sumner.

Ironically, for all his influence and expertise, Dellenbaugh never saw Separation Rapid. The second Powell party abandoned the river at Kanab Creek, approximately halfway through Grand Canyon. The Howlands and Dunn cruised farther through Grand Canyon than Dellenbaugh.

Powell created the 1875 diary narrative under pressure from Congress. In 1874 he went before the House Appropriations Committee in order to obtain government funds to continue his land surveys. He was told appropriations would be granted only if he promised to write an account of his explorations, to be published by the government. He made the promise, and set about the task.

Powell took four years' work involving two river trips and scores of men, and created a 97-day diary. Besides his imagination he used all the material he could gather. This included his own diaries from the first and second trips, his

1869 letters to the *Chicago Tribune,* Sumner's diary, Walter Powell's letter to the *Chicago Evening Journal,* and Almon Harris Thompson's diary of the second expedition. Congress, impressed, but unaware of the manipulation of records and facts, voted the needed funds. Powell's *Report on Exploration of the Colorado River of the West and its Tributaries* was published by the U.S. Government Printing Office in 1875. As a result of the Report, Powell's stature increased. He convinced Congress and the public the creation was his diary. He may have also convinced himself, for he never once publicly or privately admitted that it was a compilation of the four years' work.

Obviously Powell lied when he told Stanton the diary in his 1875 Report was "written on the spot." He lied when he defended its incidents as not being exaggerated. Stanton was to find this out later when he discovered Powell's original 1869 diary that was "written on the spot." Unaware of Powell's manipulation of other records, Stanton came to several erroneous conclusions. The first part of the voyage, from Green River, Wyoming, to the mouth of the Uinta, was not written from memory (page 123). It was based on his letters to the *Chicago Tribune* plus several other documents. Powell never tried to hide the second trip. On the contrary, at his urging several members of the expedition wrote letters to their home newspapers. It appears Stanton was not aware of these letters.

Stanton was correct when he accused Powell of manipulating the facts. When he came to write an account of Powell's first trip, Stanton scrutinized the Report "diary" and began to weed out the inaccuracies and absurdities. From his own experience he knew many of the descriptions of rapids and canyon walls were highly exaggerated, and struggled to set the record straight. When he began his correspondence with Sumner and Hawkins, Stanton's major purpose was to determine what led up to the break in the party, which ended with the separation affair. As he had been with Dellenbaugh, Sumner was reluctant to open up. He referred Stanton to Hawkins and said, "he (Hawkins) will say things that I don't like to say." It was not until Stanton convinced him the reputations of the Howlands and Dunn were at stake that Sumner decided to give his version. Both Sumner and Hawkins agreed the three men should not have been labeled cowards and deserters.

It is important to note there was no collusion between Sumner and Hawkins. They wrote their versions at different times from different places.

Stanton edited Sumner's account. In several places "hell" and "damn" were omitted and replaced with gentler terms. He also added the title of "Major" before Powell's name. In other places he sought to clarify certain locations as when Sumner wrote, "A few miles below PaReh

River . . ." Stanton changed it to read, "Near the center of Marble Canyon . . ."

A few strong personal comments not connected with the river trip were omitted. On page 181 Sumner's account originally read, "I presume Kelly's Hole is now ruined by some damn Mormon sheepherders." Stanton saw fit to omit the words, "damn Mormon." Also on that same page, at the end of the third paragraph, Sumner had written: "There is much good agriculture land in the Uinta Valley but it has been set aside as a indian reservation. It is hard to say when it will be settled and improved. Certainly not as long as the cattle barons can sap settlers out and use the range for their own selfish purposes." The entire remark was omitted by Stanton in the final manuscript.

Another comment left out by Stanton involved gunplay at one of the camps. On page 202 after the sentence, "He did not accept the proposition . . ." Sumner had written, ". . . which so angered Hall that he wanted to blow the top of his damn head off, and I believe he would have done so if Hawkins had not disarmed him." This surprising omission coincides with Hawkins's version to Stanton on page 148 and to Bass, pages 159, 160.

One should realize when reading the Sumner and Hawkins accounts that they were written almost forty years after the event to clear the names of friends falsely accused of cowardice. Many of the statements may raise eyebrows, only because this side of Powell's character has been so little known. It is difficult to conclude whether the statements are factual. Did Sumner talk Powell into exploring the Colorado River? There is no record of Powell's intention to do so before meeting Sumner. Was Powell domineering with the crew? Both accounts agree that he was.

When Dellenbaugh read Sumner's account, he immediately resorted to name-calling. "Sumner was sore at Powell." Later, "Hawkins was an ignoramus" and "Bass had a grudge against Powell." Anyone who was uncomplimentary of Powell was automatically characterized as being against him. There are still many Dellenbaugh-type writers and historians around. They would rather think of Powell as a knight in shining armor than a mere mortal. He had his faults; to develop a true picture they cannot be overlooked.

Until publication of *Colorado River Controversies* in 1932, history dealt only with the heroic exploits of John Wesley Powell. There has always been this other side to the story that has been conveniently overlooked because it was considered too controversial.

There is still much to be learned about the Powell voyages and survey.

MARTIN J. ANDERSON

Jersey City
July 1979

Martin Anderson became interested in the Powell expeditions as a result of his encounters with the late Otis "Dock" Marston, noted authority on river running in the Grand Canyon. Campfire yarning and speculation during two Marston-guided trips in 1972-73 prompted Anderson to begin probing primary source materials. His article, "First Through the Canyon, Powell's Lucky Voyage in 1869," appeared in the Winter 1979 issue of **The Journal of Arizona History.**